Polymer Synthesis and Characterization

Polymer Synthesis and Characterization
A LABORATORY MANUAL

STANLEY R. SANDLER
Elf Atochem North America, Inc.
King of Prussia, Pennsylvania

WOLF KARO
Consultant to the Chemical Industry
Huntington Valley, Pennsylvania

JO-ANNE BONESTEEL
Elf Atochem North America, Inc.
King of Prussia, Pennsylvania

ELI M. PEARCE
Polytechnic University
Brooklyn, New York

ACADEMIC PRESS
San Diego   London   Boston   New York
Sydney   Tokyo   Toronto

Cover illustration: GPC column separation of a polymer (see Experiment 18, Figure 1)

This book is a guide to provide general information concerning its subject matter: it is not a procedural manual. Synthesis of chemicals is a rapidly changing field. The reader should consult current procedural manuals for state-of the-art instructions and applicable government safety regulations. The Publisher and the authors do not accept responsibility for any misuse of this book, including its use as a procedural manual or as a source of specific instructions. The users of this book knowingly and voluntarily assume all the risks of any and all injuries that may result from performing any experiment herein.

This book is printed on acid-free paper.

Copyright © 1998 by ACADEMIC PRESS

All Rights Reserved.
No part of this publication may be reproduced or transmitted in any form or by any means, electronic or mechanical, including photocopy, recording, or any information storage and retrieval system, without permission in writing from the publisher.

Academic Press
a division of Harcourt Brace & Company
525 B Street, Suite 1900, San Diego, California 92101-4495, USA
http://www.apnet.com

Academic Press Limited
24-28 Oval Road, London NW1 7DX, UK
http://www.hbuk.co.uk/ap/

International Standard Book Number: 0-12-618240-X

PRINTED IN THE UNITED STATES OF AMERICA
98 99 00 01 02 03 EB 9 8 7 6 5 4 3 2 1

Contents

Preface xvii

Introduction 1
    SCALE OF OPERATIONS AND MONITORING OF REACTIONS 1
    SAFETY 2
    WASTE DISPOSAL 2
    LABORATORY RECORDS/NOTEBOOKS 2

SECTION I
Polymer Synthesis

A: POLYMERIZATION OF STYRENE

    INTRODUCTION 7

EXPERIMENT 1
Preparation of Polystyrene by a Free Radical Polymerization Process 9
    INTRODUCTION 9
    APPLICABILITY 9
        Bulk Polymerization 10
    SAFETY PRECAUTIONS 11
    APPARATUS 11
    REAGENTS AND MATERIALS 11
    PROCEDURE 11

NOTES 12
REFERENCES 12

EXPERIMENT 2
Preparation of Polystyrene by an Emulsion Polymerization Process 14
INTRODUCTION 14
APPARATUS 14
REAGENTS AND MATERIALS 15
SAFETY PRECAUTIONS 15
PROCEDURE 15
NOTE 15
REFERENCES 15

EXPERIMENT 3
Preparation of Polystyrene by an Anionic Polymerization Method 17
INTRODUCTION 17
APPLICABILITY 19
SAFETY PRECAUTIONS 19
APPARATUS 19
REAGENTS AND MATERIALS 20
PROCEDURE 20
REFERENCES 20

EXPERIMENT 4
Preparation of Polystyrene by a Cationic Polymerization Process 22
INTRODUCTION 22
APPLICABILITY 22
SAFETY PRECAUTIONS 23
APPARATUS 23
REAGENTS AND MATERIALS 24
PROCEDURE 24
NOTES 24
REFERENCES 25

B: POLYMERIZATION OF ACRYLIC ESTERS

INTRODUCTION 27
REACTANTS AND REACTION CONDITIONS 27
    Inhibitors and Their Removal 27
    Reaction Temperature 28

Chain-Transfer Agents   28
Initiators   28
PROCEDURES   29
Bulk Polymerization   29
Suspension Polymerization   30
Emulsion Polymerization   31
REFERENCES   32

EXPERIMENT 5
Bulk Polymerization of Methyl Methacrylate: A Test Tube Demonstration   35

INTRODUCTION   35
SAFETY PRECAUTIONS   35
APPARATUS   36
REAGENTS AND MATERIALS   36
PROCEDURE   36
NOTES   36
REFERENCES   37

EXPERIMENT 6
Suspension Polymerization of Methyl Methacrylate   38

INTRODUCTION   38
SAFETY PRECAUTIONS   38
APPARATUS   38
REAGENTS AND MATERIALS   39
PROCEDURE   39
NOTES   39
REFERENCES   40

EXPERIMENT 7
Redox Emulsion Polymerization of Ethyl Acrylate   41

INTRODUCTION   41
SAFETY PRECAUTIONS   41
APPARATUS   41
REAGENTS AND MATERIALS   42
PROCEDURE   42
NOTES   43
REFERENCES   43

C: POLYAMIDES

INTRODUCTION   45
REFERENCES   47

EXPERIMENT 8
Preparation of Poly(hexamethylenesebacamide) (Nylon 6-10) by an Interfacial Polymerization Technique 49
- INTRODUCTION 49
- BACKGROUND 49
- SAFETY PRECAUTIONS 49
- APPARATUS 50
- REAGENTS AND MATERIALS 50
- PROCEDURE 50
- NOTES 50
- REFERENCES 51

D: POLYESTERS

- INTRODUCTION 53
- REFERENCES 56

EXPERIMENT 9
Preparation of Poly(1,4-butylene isophthalate) 59
- INTRODUCTION 59
- SAFETY PRECAUTIONS 59
- APPARATUS 59
- REAGENTS AND MATERIALS 59
- PROCEDURE 60
- NOTE 60
- REFERENCE 60

E: EPOXY RESINS

- INTRODUCTION 61
- SAFETY PRECAUTIONS 62
- ANALYSIS OF EPOXY RESINS 63
- CONDENSATION–ELIMINATION REACTIONS 63
  - Epoxy Compounds via Epichlorohydrin 63
- CURING–POLYMERIZATION REACTIONS OF EPOXY COMPOUNDS AND RESINS 64
- REFERENCES 66

EXPERIMENT 10
Preparation of a Cured Epoxy Resin by the Room Temperature Reaction of Bisphenol A Diglycidyl Ether with Polyamines 69
- SAFETY PRECAUTIONS 69

APPARATUS 69
REAGENTS AND MATERIALS 69
PROCEDURE 70
REPORT 70
NOTES 70

F: POLYMERIZATION OF VINYL ACETATE

INTRODUCTION 71

EXPERIMENT 11
Seeded Emulsion Terpolymerization of Vinyl Acetate, Butyl Acrylate, and Vinyl Neodecanoate with Gradual Monomer and Initiator Additions 73
PRINCIPLE 73
COMPOSITION 74
SAFETY PRECAUTIONS 74
APPARATUS 74
REAGENTS AND MATERIALS 75
PROCEDURE 75
REPORT 76
REFERENCE 76

EXPERIMENT 12
Preparation of Poly(vinyl alcohol) by the Alcoholysis of Poly(vinyl acetate) 77
INTRODUCTION 77
BACKGROUND 77
APPLICABILITY 78
SAFETY PRECAUTIONS 78
APPARATUS 78
REAGENTS AND MATERIALS 79
PROCEDURE 79
NOTE 79
REFERENCES 80

SECTION II
Polymer Characterization

EXPERIMENT 13
Nuclear Magnetic Resonance 83
INTRODUCTION 83
PRINCIPLES 83
   Theory 83
      Polymer Characterization 86

ACCURACY AND PRECISION   91
SAFETY PRECAUTIONS   92
APPARATUS   92
REAGENTS AND MATERIALS   92
PROCEDURE   92
    Epoxy Kinetics: Condensation Reaction   93
    Polymer Tacticity   93
CALCULATIONS   95
    Epoxy Kinetics   95
    Polymer Tacticity   95
REPORT   95
NOTES   95
REFERENCES   96

EXPERIMENT 14

Infrared Spectroscopy   98

INTRODUCTION   98
BACKGROUND   98
    Electromagnetic Radiation Spectrum   98
    Theory   100
    Polymer Characterization   100
    Sample Preparation Techniques Applied to Polymers   103
APPLICABILITY   104
ACCURACY AND PRECISION   104
SAFETY PRECAUTIONS   104
APPARATUS   104
REAGENTS AND MATERIALS   105
PROCEDURE   105
FUNDAMENTAL EQUATION   105
CALCULATIONS   105
REPORT   106
REFERENCES   106

EXPERIMENT 15

Thermogravimetric Analysis   108

INTRODUCTION   108
PRINCIPLE   108
    Qualitative Identification   109
    Compositional Analysis   109
APPLICABILITY   112
ACCURACY AND PRECISION   112
SAFETY PRECAUTIONS   112
APPARATUS   112
REAGENTS AND MATERIALS   112
PROCEDURE   113

Experiment 15a: Polymer Fingerprinting   114
    PURPOSE   114
Experiment 15b: Polymer Degradation in Different Atmospheres   115
    FUNDAMENTAL EQUATIONS   116
    CALCULATIONS   116
    REPORT   117
    NOTES   117
    REFERENCES   118

EXPERIMENT 16
Differential Scanning Calorimetry   120
    INTRODUCTION   120
    BACKGROUND   120
        Theory   120
        Heat Capacity   122
        Enthalpy   122
        Glass Transition   122
        Melting and Crystallization   123
        Polymerization and Heats of Reaction   124
    APPLICATIONS   124
    ACCURACY AND PRECISION   124
    SAFETY PRECAUTIONS   125
    APPARATUS   125
    REAGENTS AND MATERIALS   125
    PREPARATION   125
    PROCEDURE   125
        First-Order Transition of Poly(amide)   126
    FUNDAMENTAL EQUATIONS   127
    CALCULATIONS   127
    REPORT   128
    NOTES   128
    REFERENCES   128

EXPERIMENT 17
Dilute Solution Viscosity of Polymers   131
    INTRODUCTION   131
    PRINCIPLE   131
        Flexible Chains   133
        Evaluation of Molecular Weight   133
        Branched Polymers   134
        Copolymers   134
        Polyelectrolytes   134
    APPLICABILITY   134
    ACCURACY AND PRECISION   134
    SAFETY PRECAUTIONS   134

APPARATUS 135
REAGENTS AND MATERIALS 135
PREPARATION 135
PROCEDURE 136
FUNDAMENTAL EQUATIONS 137
CALCULATIONS 137
REPORT 138
NOTES 138
REFERENCES 138

EXPERIMENT 18
Gel Permeation Chromatography 140
INTRODUCTION 140
THEORY 140
    Background 140
    Size Exclusion 141
    Column Efficiency 143
    Column Packings 143
    Eluents 144
    Detectors 144
APPLICABILITY 145
ACCURACY AND PRECISION 146
SAFETY PRECAUTIONS 146
APPARATUS 146
REAGENTS AND MATERIALS 146
PREPARATION 147
PROCEDURE 148
    Instrument Setup 148
FUNDAMENTAL EQUATIONS 148
CALCULATIONS 149
REPORT 149
NOTES 150
REFERENCES 151

EXPERIMENT 19
Light Scattering 152
INTRODUCTION 152
THEORY 152
    Background 152
    Light Scattering from a Liquid 153
    Scattering from a Solution of Small Particles 153
    Scattering from a Solution of Larger Particles 154
    Treatment of Data: General Equation and Zimm Plot 154
    Instrumentation 155

Contents   **xiii**

    APPLICABILITY   156
    ACCURACY AND PRECISION   156
    SAFETY PRECAUTIONS   156
    APPARATUS   157
    REAGENTS AND MATERIALS   157
    PREPARATION   157
    PROCEDURE   158
        Calibration of Light Scattering Photometer   159
        Measuring the Scattering of the Pure Solvent   159
        Measuring the Scattering of the Reference Material   159
        Measuring the Scattering of the Polymer Solutions   159
    FUNDAMENTAL EQUATIONS   159
    CALCULATIONS   160
    REPORT   160
    NOTES   160
    REFERENCES   161

EXPERIMENT 20
End Group Analysis   163

    INTRODUCTION   163
    THEORY   163
        Molecular Weight Determination   163
        Linear Polymers   164
        Condensation Polymers   164
        Vinyl Polymers   164
        Other Functional Groups   165
        Methodology   166
    APPLICABILITY   166
    ACCURACY AND PRECISION   166
    SAFETY PRECAUTIONS   166
    APPARATUS   166
        Procedure I: Amine End Groups   166
        Procedure II: Hydroxyl End Groups   167
    REAGENTS AND MATERIALS   167
        Procedure I: Amine End Groups   167
        Procedure II: Hydroxyl End Groups   167
    SAMPLE PREPARATION   167
        Procedure I: Amine End Groups   167
        Procedure II: Hydroxyl End Groups   168
    PROCEDURE   168
        Procedure I: Amine End Groups   168
        Procedure II: Hydroxyl End Groups   169
    FUNDAMENTAL EQUATIONS   170
        Procedure I: Amine End Groups   170
        Procedure II: Hydroxyl End Groups   170

CALCULATIONS 170
    Procedure I: Amine End Groups 170
    Procedure II: Hydroxyl End Groups 170
REPORT 170
    Procedure I: Amine End Groups 170
    Procedure II: Hydroxyl End Groups 171
NOTE 171
REFERENCES 171

EXPERIMENT 21
X-Ray Diffraction 173

INTRODUCTION 173
PRINCIPLE 173
    Theory 173
    Amorphous Samples 175
    Degree of Orientation 175
    Crystalline Samples 176
    Experimental Methods 177
APPLICABILITY 178
ACCURACY AND PRECISION 178
SAFETY PRECAUTIONS 178
APPARATUS 179
REAGENTS AND MATERIALS 179
PREPARATION 180
    Sample Mounting 180
    Fiber Preparation 180
PROCEDURE 180
    Start-up 180
    Wide Angle Experiments 180
    Small Angle Experiments 181
FUNDAMENTAL EQUATIONS 181
CALCULATIONS 181
REPORT 182
NOTES 183
REFERENCES 183

EXPERIMENT 22
Optical Microscopy 185

INTRODUCTION 185
BACKGROUND 185
    Polymer Morphology: Single Crystals 185
    Theory 186
    Imaging Modes 188
    Polarized Light 188
    Measurement of Refractive Index 190

Dynamic Microscopy   191
Sample Preparation   191
APPLICABILITY   192
ACCURACY AND PRECISION   192
SAFETY PRECAUTIONS   192
APPARATUS   192
REAGENTS AND MATERIALS   193
PREPARATION   193
PROCEDURE   193
Microscope Setup   193
Crystallization and Melting of Linear Poly(ethylene)   193
Birefringence   194
FUNDAMENTAL EQUATIONS   195
CALCULATIONS   195
REPORT   196
NOTES   196
REFERENCES   196

EXPERIMENT 23
Dynamic Mechanical Analysis   198

INTRODUCTION   198
PRINCIPLE   198
The Modulus Curve   198
Theory   199
Amorphous Polymers   201
Crystalline Polymers   202
Elastomers   203
APPLICABLILITY   203
ACCURACY AND PRECISION   203
SAFETY PRECAUTIONS   203
APPARATUS   203
REAGENTS AND MATERIALS   204
PREPARATION   204
PROCEDURE   204
Calibration   204
Measurement   204
FUNDAMENTAL EQUATIONS   205
CALCULATIONS   205
REPORT   205
NOTES   205
REFERENCES   206

Index   207

Preface

This laboratory manual contains some of the procedures from the three-volume set of *Polymer Syntheses,* Second Edition, by Stanley R. Sandler and Wolf Karo, as well as new procedures and a new section on polymer characterization involving viscosity, GPC (SEC), NMR, IR, TGA, DSC, and other techniques, written by Dr. Jo-Anne Bonesteel. Professor Eli M. Pearce also joined our team to develop a more timely and relevant polymer laboratory manual, and his co-authors are grateful for both his advice and comments and his checking of the procedures.

We also acknowledge the assistance of the various checkers for their comments and suggestions that were incorporated into the manual. We thank Shaoxiang Lu, Ping-Tsung Huang, Yukai Han, and Shih-Chien Chiu from Polytechnic University; Richard Perrinaud, Dana Garcia, Larry Judovits, Nafih Mekhilef, Marina Despotopoulou, Tuandung Nguyen, Russell Lewis, Herminder Sidhu, and Robert Carter from Elf Atochem North America; and Kathy Lavanga from Rheometric Scientific, Inc.

Finally, we thank Ms. Shelley Knepp and Ms. Erin Torre for their help in typing the manuscript. In addition, we thank our families for their support and understanding for the many hours we spent immersed in paperwork during evenings and weekends in preparing the manuscript.

Stanley R. Sandler
Wolf Karo
Jo-Anne Bonesteel
Eli M. Pearce

Introduction

This manual has been designed as an up-to-date polymer laboratory manual useful for both students and industrial chemists.

In the case of students, this laboratory manual provides examples of the synthesis of the major classes of polymers along with a separate section on polymer characterization experiments and techniques widely used by industrial researchers. Most of the preparations and characterization experiments have been student tested and reviewed by Professor Eli Pearce at the Polytechnic University.

Many, but not all, of the preparations have been taken from the three-volume series of "Polymer Syntheses," Volumes I, II, and III published by Academic Press in 1992, 1994, and 1996.

This laboratory manual assumes that the student is already familiar with organic chemistry and has taken a course in polymer chemistry where the mechanisms of the various polymer reactions illustrated by the preparations in this manual have already been covered. Careful record keeping is essential and is covered in a separate section later. Experience in the various analytical techniques such as infrared (IR) and nuclear magnetic resonance (NMR) is also assumed. Experience in distillation, both at atmospheric pressure and at reduced pressure, is also assumed. Where possible, monomers are used with little purification except for inhibitor removal and drying by students in order to save time. However, when careful kinetics are required, then very careful purification is a necessity.

SCALE OF OPERATIONS AND MONITORING OF REACTIONS

The majority of preparations can be scaled down provided that microware is available. Most preparations are already scaled down and can be used as described. Reactions run on a large scale by a laboratory class pose a disposal problem, which is very costly.

Polymer preparations should not be scaled up without a careful review and a gradual scaleup to check exotherms. This will determine the proper equipment and cooling needed before starting. All glassware should be free of cracks, and defects before using. In most cases, ordinary laboratory glassware may be used, but resin kettles are sometimes desirable on a larger scale operation.

Polymerization reactions can be followed by IR, by NMR, by viscosity measurements, or by other techniques (see the instructor).

SAFETY

All experiments must be carried out in a fume hood with the use of proper personal protective equipment such as safety glasses with side shields, a laboratory coat or apron, and the proper gloves for hands.

Material safety data sheets (MSDS) for each chemical being used must be read and understood with approval to go ahead obtained by the instructor. A sign-off form should be used by the instructor to make sure that the student understands the hazards and measures for protection in case of a spill.

WASTE DISPOSAL

1. No chemical or reaction product should be disposed of by pouring down a sink drain.
2. The instructor or laboratory supervisor will provide the student with the proper procedure and the location where waste is to be placed (usually a separate labeled container in a hood).
3. All broken glassware and dirty paper towels must also be segregated for separate disposal. See the instructor or supervisor for the procedure to follow.

LABORATORY RECORDS/NOTEBOOKS

It is important that the laboratory record reflect the exact stage of each procedure as it is carried out. This is essential when the experimentalist carries out several preparations simultaneously. It is even more important as a means of communication between chemists working on various phases of a single project. After interruptions or emergencies, the record helps in determining how to proceed with the work. The notes should help in detecting causes of unusual observations, as well as causes of accidents or fires. In other words, there are important safety considerations. It cannot be emphasized enough that detailed and accurate notes are essential for patent applications and patent priority claims.

Most importantly, it is a matter of scientific integrity and personal ethics that the notes reflect the experimenter's activities, observations, thoughts, and conclusions at a specific time.

In the industrial environment, so-called "good manufacturing practices" (GMP) are now standard. In part, GMP procedures resulted from various governmental regulations. It is useful to the working chemist in its utility in managing time, equipment, health, and safety, as well as in environmental and material disposal concerns.

The authors therefore strongly urge that GMP procedures be developed and inculcated from the beginning of a laboratory course. It is hoped that GMP procedures will not be another onerous matter to bedevil the student but rather a natural aspect of laboratory activity.

To be most effective, each student should have a clipboard posted near the work area. This clipboard will include all the information needed for the experiment:

1. MSDS of all materials that are in the work area, which should encourage the disposal of unused chemicals and discourage the storage of unnecessary reagents.
2. The scheduled date and time for the preparation.
3. The title of the experiment.

4. A brief summary of the experiment that is to be performed.

5. Notes or special instructions from the instructor or from classroom discussions.

6. A table of physical constants of the materials that are to be used, of the products, coproducts, and by-products.

7. Notes on the proposed disposal of the products and associated materials.

8. Preliminary approval by the instructor or laboratory assistant and signature of the experimenter.

9. A step-by-step outline of the procedure with the time and date of the actual start of the work, a space to check off that a step has been taken, the starting time of the step, the time when the step has been completed, observations and notes, if any. (For a typical example, see Table 1.)

10. Final description of the product and details (with time, date, and location) of the disposal of all materials, followed by a note on the cleanup of the work area.

11. Signature that the work has been completed by the chemist and signature with a note that the write-up has been read and understood by the instructor or other supervisor with appropriate dates and times.

12. Paper work that has been generated should be page numbered (with initials) and preserved, ultimately to be bound. The student's laboratory notebook should also summarize the work that has been done. The exact procedures that are used for the preservation of notes on experiments vary from institution to institution. The student should be instructed in the appropriate method what is used at the particular laboratory under consideration.

There is room for considerable individuality in the step-by-step outline and format of a procedure. The essential points are that each step of the operating instructions is listed and that there are provisions so that a log of the procedure is recorded.

Table 1 shows an example of the partial operating instructions and log for the preparation of a terpolymer of vinyl acetate, butyl acrylate, and vinyl neodecanoate. The details of these instructions will, of course, have to be individualized for specific situations.

TABLE 1
Operating Instructions and Log

Product: Vinyl acetate, butyl acrylate, and vinyl neodecanoate
(60/15.3/24.7[w/w]) latex polymer

Notebook reference:
Approval:
Date:
Equipment:
Materials:

Operating Instructions

Date: Time: (This must be noted after each step)

1. In a hood, set up a 1-liter four-necked flask fitted with a mechanical stirrer, two Claisen-type adapters to accommodate a reflux condenser, a 250-ml addition funnel, a nitrogen inlet tube, a thermometer, and a 50-ml addition funnel. Check off
2. Prepare the *initiator feed* by dissolving 0.4 g of ammonium persulfate in 40 ml of deionized water. Check off
3. Place the initiator feed solution in the 50-ml addition funnel. Check off
4. In a 250-ml flask, prepare the *monomer feed* solution by dissolving
   44.4 g of vinyl neodecanoate and Check off
   27.6 g of butyl acrylate Check off
   in 108 g of vinyl acetate. Check off
5. Place 136 ml of deionized water in the 1-liter four-necked flask. Check off
6. In turn, using the mechanical stirrer at as low speed as possible, dissolve
   4.00 g of Cellosize hydroxyethyl cellulose WP-300, Check off
   2.00 g of Tergitol nonionic surfactant NP-40, Check off
   2.60 g of Tergitol nonionic surfactant NP-15, Check off
   2.20 g of Alcolac Siponate DS-4, Check off
   and 0.40 g of ammonium bicarbonate in deionized water. Check off
7. To the flask add 12 g of vinyl acetate, Check off
   3.00 g of butyl acrylate, Check off
   and 5.00 g of vinyl neodecanoate. Check off
8. Turn on the stirrer (run at about 150–200 rpm), blanket the flask contents with nitrogen, and heat the flask to 55°C. Check off
9. Add 0.16 g of ammonium persulfate and attach the two addition funnels to the reaction flask. Check off
10. Maintain the reaction temperature at 55°C for 20 min and then raise it to 75°C. Check off
11. While the temperature in the reaction flask is being raised, place the monomer feed solution in the 250-ml addition funnel and Check off
    the initiator solution in the 50-ml addition funnel (as per step 3). Check off
12. After maintaining the reaction temperature at 75°C for 15 min, raise the temperature to 78°C. Check off
13. Start the addition of the monomer feed and the initiator feed at reasonably uniform rates so that the monomer addition takes 120 min and the initiator addition takes 150 min while maintaining a reaction temperature of 76–80°C.

Record the reaction history in tabular form:

Time	Temp.	Monomer vol.	Initiator vol.	Observations

(Then continue with steps 14 and 15)

14. After the additions have been completed, continue heating and stirring for an additional hour. Check off
15. Cool and filter the latex that has been produced. Determine the yield, pH, and percent solids. Check off

APPROVAL
Prepared by: Date:
Read and understood by: Date:
Notebook reference:

SECTION

Polymer Synthesis

POLYMERIZATION OF STYRENE

INTRODUCTION

The following experiment describes the background of free radical polymerizations of vinyl monomers. This information will be experimentally illustrated in experiments involving bulk and emulsion polymerizations of styrene.

The mechanisms of polymerization will not be discussed here but several worthwhile references should be consulted [1–14]. This section gives mainly examples of some selected preparative methods for carrying out the major methods of polymerization as encountered in the laboratory. All intrinsic viscosities listed in this section have units of dl/g.

EXPERIMENT 1

Preparation of Polystyrene by a Free Radical Polymerization Process

INTRODUCTION

In 1838 Regnault [15] reported that vinylidene chloride could be polymerized. In 1839 Simon [16] and then Blyth and Hofmann (1845) [17] reported the preparation of polystyrene. These were followed by the polymerization of vinyl chloride (1872) [18], isoprene (1879) [19], methacrylic acid (1880) [20], methylacrylate (1880) [21], butadiene (1911) [22], vinyl acetate (1917) [23], vinyl chloroacetate [23], and ethylene (1933) [24]. Klatte and Rollett [23] reported that benzoyl peroxide is a catalyst for the polymerization of vinyl acetate and vinyl chloroacetate.

In 1920 Staudinger [25] was the first to report on the nature of olefin polymerizations leading to high polymers. A great many of his studies were carried out on the polymerization of styrene. These studies led to recognition of the relationship between relative viscosity and molecular weight [26,27]. The radical nature of these reactions was later elucidated by Taylor [28], Paneth and Hofeditz [29], and Haber and Willstätter [30]. The understanding of the mechanism of polymerization was greatly aided by Kharasch et al. [31], Hey and Waters [32], and Flory [4,33].

No effort will be made to discuss the mechanism of polymerization, but let it suffice to say that the polymerization is governed by the steps shown in Eqs. (2) and (3), in (4), (5), and (6), and in (8).

APPLICABILITY

The most common initiators are acyl peroxides, hydroperoxides, or azo compounds. Hydrogen peroxide, potassium persulfate, and sodium perborate are popular in aqueous systems. Ferrous ion in some cases enhances the catalytic effectiveness.

Ethylene is conveniently polymerized in the laboratory at atmospheric pressure using a titanium-based coordination catalyst [34]. It may also be polymerized less conveniently in the laboratory under high pressures using free radical catalysts at high and low temperatures [35–37]. Other olefins such as propylene, 1-butene, or 1-pentene homopolymerize free radically only to low molecular weight polymers and require ionic or coordination catalysts to afford high molecular weight polymers [38–41]. These olefins can effectively be copolymerized free radically.

The free radical polymerization process, which can be carried out in the laboratory, is best illustrated by the polymerization of styrene.

Free radical polymerization processes [41] are carried out in bulk, solution, suspension, emulsion, or by precipitation techniques. In all cases the monomer used should be free of solvent and inhibitor or else a long induction period will result. In some cases this may be overcome by adding excess initiator.

Initiator:
$$I_2 \longrightarrow 2I\cdot \tag{1}$$

Initiation:
$$CH_2=CHR + I\cdot \longrightarrow I-CH_2-\underset{R}{CH}\cdot \tag{2}$$

Propagation:
$$I-\underset{R}{CH_2CH}\cdot + CH_2=CHR \longrightarrow I-CH_2-\underset{R}{CH}-CH_2-\underset{R}{CH}\cdot \xrightarrow{nCH_2=CHR}$$
$$I(CH_2CH)_{n+1}\underset{R}{CH_2CH}\cdot \tag{3}$$

Termination (by radical coupling, disproportionation, or chain transfer):

Radical coupling:
$$\sim\sim CH_2\underset{R}{CH}\cdot + \sim\sim CH_2-\underset{R}{CH}\cdot \longrightarrow \sim\sim CH_2\underset{R}{CH}\underset{R}{CH}CH_2\sim\sim \tag{4}$$

Disproportionation of two radicals:
$$\sim\sim CH_2-\underset{R}{CH}\cdot + \sim\sim CH_2-\underset{R}{CH}\cdot \longrightarrow \sim\sim CH_2\underset{R}{CH_2} + \sim\sim CH=\underset{R}{CH} \tag{5}$$

Chain transfer:
$$\sim\sim CH_2\underset{R}{CH}\cdot + R'SH \longrightarrow \sim\sim CH_2\underset{R}{CH_2} + R'S\cdot \tag{6}$$

$$R'S\cdot + CH_2=\underset{R}{CH} \longrightarrow RS'CH_2\underset{R}{CH}\cdot \quad \text{(start of new monomer chain)} \tag{7}$$

Bulk Polymerization

Bulk polymerization consists of heating the monomer without solvent with initiator in a vessel. The monomer–initiator mixture polymerizes to a solid shape fixed by the shape of the polymerization vessel. The main practical disadvantages of this method are the difficulty in the removal of polymer from a reactor or flask and the dissipation of the heat evolved by the polymerization.

In the use of polystyrene, the polymerization reaction is exothermic to the extent of 17 Kcal/mol or 200 BTU/lb (heat of polymerization). The polystyrene produced has a broad molecular weight distribution and poor mechanical properties. The residual monomer in the ground polymers can be removed using efficient devolatilization equipment. Several reviews are worthwhile consulting [42–44].

The bulk polymerization of styrene to give a narrow molecular weight distribution has appeared in a U.S. patent [45]. The polydispersity reported was

2.6 at a 93% conversion and had an average molecular weight of about 100,000. This was accomplished by polymerizing styrene in the presence of 1.0% of 4-*tert*-butylpyrocatechol at 127°C for 2.27 hr. Heating in the absence of the latter gave a polydispersity of 3.3 with an average molecular weight of 79,200.

Several references to the bulk polymerization of styrene are worth consulting [46–50]. Most consider a continuous bulk polymerization apparatus with some using spraying of the monomer through a nozzle. The controlled evaporation of unreacted monomer is one method of removing the heat of reaction.

SAFETY PRECAUTIONS

Before this experiment is carried out, the student must read the material safety data sheets (MSDS) for all the chemicals used as well as for the products. The instructor must approve that you have read and understood the MSDS for the safe handling of these materials.

Please be advised that all chemicals should be considered hazardous and should be handled in a hood and with proper personal protective equipment (lab coat, proper gloves, approved safety glasses, and/or goggles). Avoid inhaling vapors and/or aerosolized materials. Avoid skin/eye contact with all chemicals at all times. Wash hands frequently. See the instructor if you have any questions or concerns.

APPARATUS

1. Balance
2. Heavy-walled polymer tube
3. Stopper for polymer tube
4. Buchner funnel
5. Filtration flask
6. Filter paper

REAGENTS AND MATERIALS

1. Styrene
2. 25% aqueous sodium hydroxide
3. Molecular sieves drying agent
4. Benzoyl peroxide
5. Azobisisobutyronitrile
6. Toluene
7. Methanol

PROCEDURE

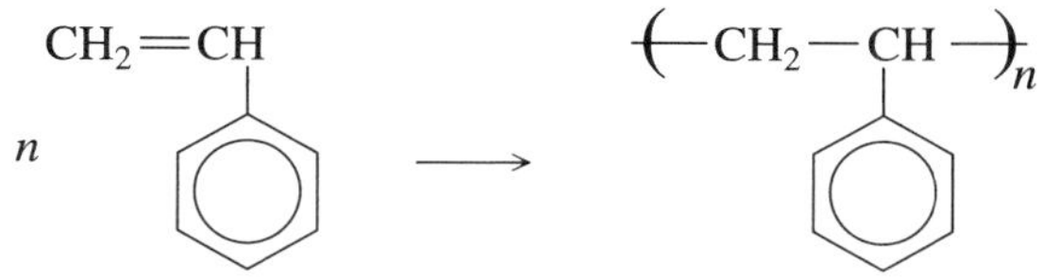

$$n\ CH_2{=}CH{-}C_6H_5 \longrightarrow {-}(CH_2{-}CH(C_6H_5)){-}_n \qquad (8)$$

1. Add 50 g of inhibitor-free (see Note 1), dry styrene (see Note 2) to a test tube.

2. Flush the tube with nitrogen and add 1.0 g of benzoyl peroxide. Gently shake, stopper, and place the tube in a oil bath at 80°C for 1–2 hr.
3. When the styrene becomes syrupy and viscous and before it stops flowing, dissolve the contents in 50 ml of toluene and then pour into 500 ml of methanol in order to precipitate the polystyrene that is formed.
4. Isolate the polymer by filtration and dry in order to calculate the percentage conversion after this time of reaction.

NOTES

1. Wash the styrene monomer twice with 25-ml portions of 25% aqueous sodium hydroxide in order to remove the inhibitor. Then wash twice with 25 ml of distilled water to remove any residual caustic reagent.

2. Dry the styrene monomer using molecular sieves and store immediately with a nitrogen blanket in a refrigerator. Although moisture will affect the reaction, the influence is minor. Dry styrene is not really necessary for free radical polymerization.

REFERENCES

1. C. Walling, "Free Radicals in Solution." Wiley, New York. 1957.
2. J. C. Bevington, "Radical Polymerization," Academic Press, New York, 1961.
3. P. E. M. Allen and P. H. Plesch, in "The Chemistry of Cationic Polymerization" (P. H. Plesch, ed.), Macmillan, New York, 1963.
4. P. J. Flory, "Principles of Polymer Chemistry," Cornell Univ. Press, Ithaca, NY, 1953.
5. R. W. Lenz, "Organic Chemistry of Synthetic High Polymers," Wiley (Interscience), New York, 1967.
6. C. L. Arcus, *Progr. Stereochem.* **3,** 264 (1962).
7. M. L. Huggins, G. Natta, V. Derreux, and H. Mark, *J. Polym. Sci.* **56,** 153 (1962).
8. G. Natta, L. Porri, P. Corradini, G. Zanini, and F. Ciampelli, *ibid. J. Polym. Sci.* **51,** 463 (1961).
9. N. Beredjick and C. Schuerch, *J. Am. Chem. Soc.* **78,** 2646 (1956).
10. C. S. Marvel and R. G. Wollford, *J. Org. Chem.* **25,** 1641 (1960).
11. G. B. Butler and R. W. Stackman, *J. Org. Chem.* **25,** 1643 (1961).
12. M. Farina, M. Peraldo, and G. Natta, *Angew. Chem. Int. Ed. Engl.* **4,** 107 (1965).
13. J. E. McGrath, *J. Chem. Ed.* **58,** 844 (1981).
14. American Chemical Society Course on "Polymer Chemistry: Principles and Practice" at Virginia Tech., Blacksburg, Virginia, March, August, and December, 1989 (Instructors: Professors M. E. McGrath, T. C. Ward, and G. L. Wilkes).
15. V. Regnault, *Ann. Chim. Phys.* **69**(2), 151 (1838).
16. E. Simon, *Ann. Pharm.* **31,** 265 (1839).
17. J. Blyth and A. W. Hofmann, *Ann. Chem. Pharm.* **53,** 289 and 311 (1845).
18. E. Baumann, *Ann. Chem. Pharm.* **163,** 312 (1872).
19. G. Bouchardat, *C. R. Acad. Sci.* **89,** 1117 (1879).
20. R. Fittig and F. Engelhorn, *Justus Liebigs Ann. Chem.* **200,** 65 (1880).
21. G. W. A. Kahlbaum, *Ber. Deut. Chem. Ges.* **13,** 2348 (1880).
22. S. W. Lebedev and N. A. Skavronskaya, *J. Russ. Phys. Chem. Soc.* **43,** 1124 (1911).
23. F. Klatte and A. Rollett, U.S. Patent 1,241,738, 1917.
24. E. W. Fawcett, British Patent 471,590, 1937.
25. H. Staudinger, *Ber. Deut. Chem. Ges.* **53,** 1073 (1920).
26. H. Staudinger, *Ber. Deut. Chem. Ges.* **62,** 241 and 292 (1929).
27. H. Staudinger, "Die Hochmolekularen Organischen Verbindungen," Springer-Verlag, Berlin/New York, 1932.
28. H. S. Taylor, *Trans. Faraday Soc.* **21,** 560 (1925).
29. F. Paneth and W. Hofeditz, *Ber. Deut. Chem. Ges.* **62,** 1335 (1929).

30. F. Haber and R. Willstätter, *Ber. Deut. Chem. Ges.* **64,** 2844 (1931).
31. M. S. Kharasch, H. Engelmann, and F. R. Mayo, *J. Org. Chem.* **2,** 288 (1937).
32. D. H. Hey and W. A. Waters, *Chem. Rev.* **21,** 169 (1937).
33. P. J. Flory, *J. Am. Chem. Soc.* **59,** 241 (1937).
34. K. Ziegler, Belgian Patents 540,459 and 543,837, 1956.
35. R. A. V. Raff and K. W. Doaks, eds., "Crystalline Olefin Polymers," Parts 1 and 2. Wiley (Interscience), New York, 1964.
36. H. V. Boening, "Polyolefins: Structure and Properties," Elsevier, New York, 1966.
37. A. Renfrew and P. Morgan, eds., "Polythene: The Technology and Uses of Ethylene Polymers," Wiley (Interscience), New York, 1960.
38. E. F. Lutz and G. M. Bailey, *J. Polym. Sci. Part A-1* **4,** 1885 (1966).
39. G. Natta, *Makromol. Chem.* **35,** 93 (1960).
40. A. Turner-Jones, *Makromol. Chem.* **71,** 1 (1964).
41. J. A. Faucher and E. P. Reding, in "Crystalline Olefin Polymers" (A. V. Raff and K. W. Doak, eds.), Chapter 13. Wiley (Interscience), New York, 1965.
42. J. L. Amos, *Polym. Eng. Sci.* **14,** 1 (1974).
43. R. F. Boyer, *J. Macromol Sci. Chem.* **15,** 1411 (1981).
44. R. H. M. Simon and D. C. Chappelear, in "Polymerization Reactions and Processes" (J. N. Henderson and T. C. Bonton, eds.), Vol. **104,** pp. 71–112, American Chemical Society, Washington, DC, 1979.
45. R. A. Hall and J. I. Rosenfeld, U.S. Pat. 4,713,421 (12/15/87); *Chem. Abstr.* **108,** 205285s (1988).
46. S. Omi, I. Iwata, K. Innbuse, M. Isu, and M. Suku, *Int. Polym. Proc.* **2** (3–4), 198 (1988); *Chem. Abstr.* **108,** 151108r (1988).
47. T. Uetake, T. Kaino, and T. Yushizawa, *Jpn. Kokai Tokkyo Koho* JP 60/152506A2 (8/10/85); *Chem. Abstr.* **104,** 110759c (1986).
48. K. T. Nguyen, E. Flaschel, and A. Renken, *Chem. Eng. Commun.* **36** (1–6), 251 (1985); *Chem. Abstr.* **103,** 24013w (1985).
49. B. M. Baysal, E. Bryramli, H. Yuruk, and B. Hazer, *Macromol. Chem.,* **6** (6) 1269 (1985); *Chem. Abstr.* **103,** 71761c (1985).
50. J. L. McCurdy, *Rev. Modern Plast.* **39** (285), 309–310, 315–316 (1980); *Chem. Abstr.* **92,** 21609o (1980).

EXPERIMENT 2

Preparation of Polystyrene by an Emulsion Polymerization Process

INTRODUCTION

The system basically consists of water and 1–3% of a surfactant (sodium lauryl sulfate, sodium dodecyl benzenesulfonate, or dodecylamine hydrochloride) and a water-soluble, free-radical generator (alkali persulfate, hydroperoxides, or hydrogen peroxide–ferrous ion). The monomer is added gradually or is all present from the start. The emulsion polymerization is usually more rapid than bulk or solution polymerization for a given monomer at the same temperature. In addition, the average molecular weight may also be greater than that obtained in the bulk polymerization process. The particles in the emulsion polymerization are of the order of 10^{-5} to 10^{-7} m in size. It is interesting to note that the locus of polymerization is the micelle and only one free radical can be present at a given time. The monomer is fed into the locus of reaction by diffusion through the water where the reservoir of the monomer is found. If another radical enters the micelle, then termination results because of the small volume of the reaction site. In other words, in emulsion polymerization the polymer particles are not formed by polymerization of the original monomer droplets but are formed in the micelles to give polymer latex particles of a very small size. For a review of the roles of the emulsifier in emulsion polymerization, see Dunn [1]. Other references with a more detailed account of the field should be consulted [2–8]. These references are only a sampling of the many that can be found in *Chemical Abstracts*.

The polymer in emulsion polymerization is isolated by either coagulating or spray drying.

APPARATUS

1. Balance
2. Resin kettle
3. Mechanical stirrer (shaft, blade, bearing, motor)
4. Condenser
5. Thermometer
6. Nitrogen inlet tube
7. Water bath/hot plate or heating mantle

REAGENTS AND MATERIALS

Chemicals

1. Styrene
2. Distilled water
3. Potassium persulfate
4. Sodium stearate
5. Sodium dodecylbenzenesulfonate
6. Sodium lauryl sulfate
7. Alum

SAFETY PRECAUTIONS

Before this experiment is carried out, the student must read the material safety data sheets (MSDS) for all the chemicals used as well as for the products. The instructor must approve that you have read and understood the MSDS for the safe handling of these materials.

Please be advised that all chemicals should be considered hazardous and should be handled in a hood and with proper personal protective equipment (lab coat, proper gloves, approved safety glasses, and/or goggles). Avoid inhaling vapors and/or aerosolized materials. Avoid skin/eye contact with all chemicals at all times. Wash hands frequently. See the instructor if you have any questions or concerns.

PROCEDURE [8]

1. Add 128.2 g of distilled water, 71.2 g of styrene, 31.4 ml of 0.680% potassium persulfate, and 100 ml of 3.56% soap solution (see Note 1) to a three-neck round bottom flask equipped with a mechanical stirrer, condenser, and nitrogen inlet tube.
2. Purge the system with nitrogen to remove dissolved air.
3. Raise the temperature to 80°C and keep there for about 3 hr to afford a good conversion of polymer.
4. Isolate the polymer by freezing–thawing or by adding alum solution and boiling the mixture.
5. Filter, wash with water, and dry the polystyrene.
6. Determine the yield based on the weight of the polystyrene isolated.

NOTE

1. It is preferable to use either sodium dodecyl benzenesulfonate or sodium lauryl sulfate (the latter is preferred).
2. For student preparations it is preferable to reduce the scale by one-tenth for all ingredients.

REFERENCES

1. A. S. Dunn, *Chem. Ind. (London)* p. 1406 (1971).
2. F. A. Bovey, I. M. Kolthoff, A. F. Medalia, and E. J. Meehan, "Emulsion Polymerization." Wiley (Interscience), New York, 1955.

3. S. N. Sautin, P. A. Kulle, and N. I. Smirnov, *Zh. Prikl Khim* (*Leningrad*) **44,** 1569 (1971).
4. A. W. De Graff, "Continuous Emulsion Polymerization of Styrene in a One Stirred Tank Reactor." Lehigh Univ. Press, Bethlehem, PA, 1970.
5. O. Gellner, *Chem. Eng.* (*New York*) **73,** 74 (1966).
6. A. G. Parts, D. E. Moore, and J. G. Waterson, *Makromol. Chem.* **89,** 156 (1965).
7. E. W. Duck, *Encycl. Polym. Sci. Technol.* **5,** 801 (1966).
8. I. M. Kolthoff and W. J. Dale, *J. Am. Chem. Soc.* **69,** 441 (1947).

EXPERIMENT 3

Preparation of Polystyrene by an Anionic Polymerization Method

INTRODUCTION

The anionic polymerization of styrene was first reported in 1914 by Schlenk and co-workers [1] and reinvestigated by Szwarc [2,3] and others [4,5]. More recently, Priddy [4,5] reported that the anionic polymerization of styrene is industrially feasible. A process to produce a broad molecular weight distribution of polystyrene ($\bar{M}_w/\bar{M}_n = >2.0$) in a continuously stirred tank reactor (CSTR) at 90–110°C has been described [6,7]. The polymers also have excellent color and are purer than those formed in the free radical polymerization process. The use of α-methylstyrene comonomer gave increased heat resistance [8]. Almost 60 years ago sodium and lithium metal were used to polymerize conjugated dienes such as butadiene [1,9,10], isoprene [9], 1-phenylbutadiene [1], and 2,3-dimethylbutadiene [11]. In 1929 Ziegler [12] described the addition of organoalkali compounds to a double bond. In 1940 the use of butyllithium for the low-pressure polymerization of ethylene was described [13]. In 1952 the kinetics of the anionic polymerization of styrene using KNH_2 was reported [14]. Some anionic polymerizations have been described as living polymers in the absence of impurities) [3,15–18].

Electron-withdrawing substituents adjacent to an olefinic bond tend to stabilize carbanion formation and thus activate the compound toward anionic polymerization [19].

The relative initiator activities are not always simple functions of the reactivity of the free anion but probably involve contributions by complexing ability, ionization, or dissociation reactions [20–25].

Waack and Doran [26] reported on the relative reactivities of 13 structurally different organolithium compounds in polymerization with styrene in tetrahydrofuran at 20°C. The reactivities were determined by the molecular weights of the formed polystyrene. The molecular weights are inversely related to the activity of the respective organolithium polymerization initiators. Reactivities decreased in the order alkyl > benzyl > allyl > phenyl > vinyl > triphenylmethyl as shown in Table 3.1.

The structure–reactivity behavior found for similar organosodium polymerization initiators of styrene [27] or that for addition reactions with 1,1-diphenylethylene [28] is almost identical with that found for the lithium initiators of Table 3.1. It is interesting to note from Table 3.1 that the reactivity of lithium

TABLE 3.1
"Standard Polymerizations" of Styrene in Tetrahydrofuran Solution at 20°C

Organolithium catalyst	Mol. wt. of polymer[a] (temp, °C)
t-Butyl	3,200[b] (−66)
	3,200 (−40)
sec-Butyl	3,500[b] (−69)
Ethyl	3,500
n-Butyl	3,600
α-Methylbenzyl	3,700
Crotyl	6,500
Benzyl	6,700
Allyl	9,600
p-Tolyl	9,900
Phenyl[c]	12,000
Phenyl[d] (LiX)	24,000 (LiCl)
	22,000 (LiBr)
Methyl	19,000
Vinyl	23,000
Triphenylmethyl[e]	66,000
Triphenylmethylsodium	53,000
Lithium naphthalene	6,000

[a] Average values.
[b] At higher temperatures there is rapid reaction with THF.
[c] Salt free.
[d] Contains equimolar lithium halide.[34]
[e] Contains equimolar LiCl. Lithium halides are indicated to have little effect on the reactivity of such resonance stabilized species. Reprinted from Ref. 26 © 1967 by the American Chemical Society. Reprinted by permission of the copyright owner.

naphthalene, a radical anion type initiator, is between that of alkyl lithiums and aromatic lithium initiators.

The anionic polymerization of styrene using the organolithium initiators can be described as a termination-free polymerization, as shown in Eqs. (1) and (2)

$$RLi + St \rightarrow RStLi \qquad (1)$$

$$RStLi + nSt \rightarrow RSt_nStLi \qquad (2)$$

The degree of polymerization is determined by the ratio of the overall rate of propagation to that of initiation.

"Living" polymers have been extensively investigated by Szwarc and co-workers [3], who have shown that the formed polymer can spontaneously resume its growth on addition of the same or different fresh monomer. Block copolymers are easily synthesized by this technique. The lack of self-termination is overcome by the addition of proton donors, carbon dioxide, or ethylene oxide.

APPLICABILITY

The living nature of the poly(styryl)anion allows one to prepare block copolymers with a great deal of control of the block copolymer structure. The preparation of diblock, triblock, and other types of multiblock copolymers has been reviewed [29–32]. Several of these block copolymers are in commercial use. The basic concept involves first preparing polystyrene block [RSt$_n$$\overline{\text{St}}$Li—see Eq. (2)] and then adding a new monomer that can be added to start another growing segment.

The living anionic ends can be functionalized by adding such agents as ethylene oxide, carbon dioxide, and methacryloyl chloride [33]. The resulting new polymer is capable of being copolymerized with additional monomers. This process can lead to the formation of various graft copolymers [29–32].

$$\begin{array}{c}
\text{H}_2\text{C}=\text{C}-\text{COCl} \\
\phantom{\text{H}_2\text{C}=\text{C}-}| \\
\phantom{\text{H}_2\text{C}=\text{C}-}\text{CH}_3
\end{array}
\quad
\begin{array}{c}
\text{RSt}_n\text{St}^- \\
\Big| \text{CO}_2
\end{array}
\quad
\begin{array}{c}
\overset{\triangledown\!\!\!\!\!\!O}{} \longrightarrow \text{RSt}_n\text{StCH}_2\text{CH}_2\text{O}^- \\
\text{CH}_3\text{OH} \\
\searrow \\
\text{RSt}_n\text{StH}
\end{array}
\tag{3}$$

$$\text{RSt}_n\text{StCOO}^-$$

$$\begin{array}{c}
\text{RSt}_n\text{StC}-\text{C}=\text{CH}_2 \\
\phantom{\text{RSt}_n\text{St}}\|| \\
\phantom{\text{RSt}_n\text{St}}\text{O}\text{CH}_3
\end{array}$$

SAFETY PRECAUTIONS

Before this experiment is carried out, the student must read the material safety data sheets (MSDS) for all the chemicals used as well as for the products. The instructor must approve that you have read and understood the MSDS for the safe handling of these materials.

Please be advised that all chemicals should be considered hazardous and should be handled in a hood and with proper personal protective equipment (lab coat, proper gloves, approved safety glasses, and/or goggles). Avoid inhaling vapors and/or aerosolized materials. Avoid skin/eye contact with all chemicals at all times. Wash hands frequently. See the instructor if you have any questions or concerns.

It is very important to keep the equipment and reaction dry as well as oxygen free during the course of this anionic polymerization process. Solvents and monomer must be delivered via a syringe and no transfers can be done by the common pouring technique.

APPARATUS

1. 100-ml Morton creased flask (three-necked)
2. Magnetic stirring apparatus
3. Magnetic stirring bar (glass enclosed)
4. Thermometer
5. Water condenser
6. Septum
7. Syringe
8. Buchner funnel

9. Buchner flask
10. Vacuum pump

REAGENTS AND MATERIALS

1. Styrene
2. Tetrahydrofuran (THF)
3. *N*-Butyllithium in hexane (2.5 *M*)
4. Methanol
5. 2-Butanone

PROCEDURE [26]

$$CH_2=CH-C_6H_5 \xrightarrow[20°C]{n\text{-BuLi}, \text{THF}} Bu-CH_2-CH^-Li^+(C_6H_5) \xrightarrow[20°C]{n\text{-}CH_2=CH(C_6H_5), \text{THF}} BuCH(C_6H_5)-CH_2-(CH(C_6H_5)-CH_2)_n Li \xrightarrow{CH_3OH} (CH(C_6H_5)-CH_2)_n \quad (4)$$

1. Connect a dry 100-ml Morton creased flask with a three-necked opening to a septum, condenser, and thermometer. Add a glass-encased magnetic stirring bar.
2. Add 6.0 ml of dry THF via a syringe and then inject 0.6 ml of 2.5 *M* *n*-butyllithium in hexane via a syringe into the flask.
3. While controlling the termperature at 20°C, add the styrene monomer (0.2 ml/sec) until 2.0 ml has been added (1.82 g or 0.0175 mol). The reaction turns a deep red color, indicating styryllithium. Polymerization is complete when all the styrene is added.
4. After 10–15 min, inject 1.0 ml of methanol to quench the reaction.
5. Precipitate the polystrene in 50 ml of cold methanol. If desired, it can be purified further by dissolving in 2-butanone and reprecipitating in methanol. When vacuum dried, the polystyrene gives approximately 1.8 g (100% yield).

REFERENCES

1. W. Schlenk, J. Appenrodt, A. Michael, and A. Thal, *Ber. Deut. Chem. Ges.* **47**, 473 (1914).
2. M. Szwarc, M. Levy, and R. Milkovich, *J. Am. Chem. Soc.* **78**, 2656 (1956).
3. M. Szwarc, "Carbanions, Living Polymers and Electron Transfer Processes," Wiley (Interscience), New York, 1968.
4. D. J. Worsfold and S. Bywater, *J. Phys. Chem.* **70**, 162 (1966).
5. F. S. Dainton *et al.*, *Makromol. Chem.* **89**, 257 (1965).

6. D. B. Priddy, U.S. Patent 4,647,632, 1987.
7. D. B. Priddy and M. Piro, U.S. Patent 4,572,819, 1986.
8. D. B. Priddy, T. D. Traugott, and R. H. Seiss, *ACS Polymer Preprints* **30**(2), Sept. 1989, p. 195. Poster Presentation at ACS National Meeting, Sept. 1989, Miami Beach, Florida.
9. C. Harries, *Justus Liebigs Ann. Chem.* **383,** 213 (1911).
10. K. Ziegler and K. Bähr, *Ber. Deut. Chem. Ges.* **61,** 253 (1928).
11. W. Schlenk and E. Bergmann, *Justus Liebigs Ann. Chem.* **479,** 42 (1930).
12. K. Ziegler, F. Crossman, H. Kliener, and O. Schafter, *Justus Liebigs Ann. Chem.* **473,** 1 (1929).
13. L. M. Ellis, U.S. Patent 2,212,155, 1940.
14. W. C. E. Higginson and N. S. Wooding, *J. Chem. Soc. London* p. 760 (1952).
15. M. Szwarc, *Nature* **178,** 1168 (1956).
16. J. Smid and M. Szwarc, *J. Polym. Sci.* **61,** 31 (1962).
17. M. Szwarc, *Encycl. Polym. Sci. Technol.* **8,** 303 (1968).
18. T. Shimomura, J. Smid, and M. Szwarc, *J. Am. Chem. Soc.* **89,** 5743 (1967).
19. D. J. Cram, *Chem. Eng. News* **41,** 92 (1963).
20. R. Waack and M. Doran, *Polymer* **2,** 365 (1961).
21. D. J. Cram, *Chem. Eng. News* **41,** 92 (1963).
22. A. A. Morton and E. Grovenstein, Jr., *J. Am. Chem. Soc.* **74,** 5434 (1952).
23. K. Yoshida and T. Morikawa, *Sci. Ind. Osaka* **27,** 80 (1953).
24. W. E. Goode, W. H. Snyder, and R. C. Fettes, *J. Polym. Sci.* **42,** 367 (1960).
25. W. H. Puterbaugh and C. R. Hauser, *J. Org. Chem.* **24,** 416 (1969).
26. R. Waack and M. A. Doran, *J. Org. Chem.* **32,** 3395 (1967).
27. A. A. Morton and E. Grovenstein, Jr., *J. Am. Chem. Soc.* **74,** 5434 (1952).
28. A. A. Morton and E. J. Lanpher, *J. Polym. Sci.* **44,** 239 (1960).
29. A. Noshay and J. McGrath, "Block Copolymers: Overview and Critical Survey," Academic Press, Orlando, FL, 1977.
30. M. J. Folkes, "Processing, Structure and Properties of Block Copolymers." Elsevier Applied Sciences, Barking, UK, 1985.
31. M. Morton, "Anionic Polymerization Principles and Practice," Academic Press, Orlando, FL, 1983.
32. L. J. Fetters, *J. Polym. Sci. Polym. Sym.* **26,** 1 (1969).
33. D. R. Iyengar and T. J. McCarthy, *ACS Polymer Preprints* **30**(2), September 1989, p. 154 (paper presented at the ACS meeting, Sept. 1989, Miami Beach, Florida).
34. R. Waack and M. A. Doran, *Chem. Ind. (London),* **496** (1964). The presence of lithium halides decreases the reactivity of phenyllithium.

EXPERIMENT 4

Preparation of Polystyrene by a Cationic Polymerization Process

INTRODUCTION

Cationic polymerization has a history dating back to the early 1800s and has been extensively investigated by Plesch [1,2], Dainton and Sutherland [3], Evans et al. [4,5], Pepper [6,7], Evans and Meadows [8], Heiligmann [9], and others [10–13]. Whitmore [14] is credited with first recognizing that carbonium ions are intermediates in the acid-catalyzed polymerizations of olefins. The recognition of the importance of proton-donor cocatalysts for Friedel–Crafts catalysts was first reported by Evans and co-workers [4,5]

$$MX_n + SH \rightleftharpoons [MX_nS]^-H^+ \quad (1)$$
$$\text{(SH and RX = Lewis base)}$$

$$MX_n + RX' \rightleftharpoons [MX_nR]^-E^+ \quad (2)$$

APPLICABILITY

Some common initiators for cationic polymerization reactions are protonic acids, Friedel–Crafts catalysts (Lewis acids), compounds capable of generating cations, or ionizing radiation.

Of all the acid catalysts used [15–17] sulfuric acid is the most common. Furthermore, sulfuric acid appears to be a stronger acid than hydrochloric acid in nonaqueous solvents. Some other commonly used catalysts are BF_3, $AlCl_3$, $SnCl_4$, $SnBr_4$, $SbCl_3$, $BeCl_3$, $TiCl_4$, $FeCl_3$, $ZnCl_2$, $ZrCl_4$, and I_2. The use of Lewis acid catalysts requires traces of either a proton donor (water) or a cation donor (a tertiary amine hydrohalide) to effectively initiate the polymerization process. For example, in the absence of the latter, rigorously dry systems cannot be used to initiate the cationic polymerization of isobutylene [18].

For alkenes, the reactivity is based on the stability of the carbonium ion formed and they follow the order tertiary > secondary > primary. Thus olefins react as follows: $(CH_3)_2C=CH_2 \simeq (CH_3)_2C=CHCH_3 > CH_3CH=CH_2 > CH_2=CH_2$. Allylic and benzylic carbonium [19,20] ions are also favored where appropriate.

The cationic polymerization process has been reviewed, and several references are worthwhile consulting [6,21,22].

Certain aluminum alkyls and aluminum dialkyl halides in the presence of proton- or carbonium-donating cocatalysts act as effective polymerization catalysts.

In the absence of monomers, trimethylaluminum (0.5 mol) reacts with t-butyl chloride (1.0 mol) at $-78°C$ to give a quantitative yield of neopentane [22]. Kennedy [23] found that aluminum trialkyls ($AlMe_3$, $AlEt_3$, $AlBu_3$) in the presence of certain alkyl halides are efficient initiators for the cationic polymerization of isobutylene, styrene, etc.

$$AlR_3 + R'X \longrightarrow [AlR_3X]^- + [R']^+ \qquad (3)$$

All experiments are carried out under a nitrogen atmosphere in stainless-steel equipment [24].

A typical example of the experimental conditions for the cationic polymerization of various olefinic and diolefinic monomers is illustrated in the Procedure section.

It should be recognized from the results that cationic polymerizations are usually initiated at low temperatures in order to suppress chain-terminating reactions and also to keep the reaction from becoming explosive in nature. These low temperatures thus favor high molecular weight polymer formation.

Substituted olefins that are capable of forming secondary or tertiary carbonium ion intermediates polymerize well by cationic initiation, but are polymerized with difficulty or not at all free radically. In general, vinyl or l-alkenes that contain electron donating groups (alkyl, ether, etc) polymerize well via a carbocationic mechanism.

SAFETY PRECAUTIONS

Before this experiment is carried out, the student must read the material safety data sheets (MSDS) for all the chemicals used as well as for the products. The instructor must approve that you have read and understood the MSDS for the safe handling of these materials.

Please be advised that all chemicals should be considered hazardous and should be handled in a hood and with proper personal protective equipment (lab coat, proper gloves, approved safety glasses, and/or goggles). Avoid inhaling vapors and/or aerosolized materials. Avoid skin/eye contact with all chemicals at all times. Wash hands frequently. See the instructor if your have any questions or concerns.

APPARATUS

1. Balance
2. Two large test tubes
3. Septums for test tubes
4. Hypodermic syringe
5. Sintered glass filter
6. Filtration flask
7. Ice water bath
8. Vacuum pump
9. Erlenmeyer flask (250 ml)

REAGENTS AND MATERIALS

1. Styrene
2. 25% aqueous sodium hydroxide
3. Molecular sieve drying agent
4. Methylene chloride
5. Stannic chloride ($SnCl_4$)
6. Methanol

PROCEDURE

$$SnCl_4 + H_2O \longrightarrow H^+(SnCl_4OH)^-$$

$$\underset{\text{PhCH=CH}_2}{} \xrightarrow{H^+(SnCl_4OH)^-} \underset{\text{PhCH}_3CH^+(SnCl_4OH)^-}{} \xrightarrow[\text{2. CH}_3OH]{\text{1. } n \text{ Styrene}} CH_3 \cdot CH\text{-Ph} (CH_2-CH\text{-Ph})_n$$

1. Fit two large previously dried test tubes with septums and a needle to act as a pressure relief valve. To one add 40 ml of dry methylene chloride via a hypodermic syringe and to the other add 30 ml of methylene chloride.
2. Place both tubes in an ice/water bath to lower the temperature to 0°C.
3. Injected 0.9 ml (1.98 g or 0.000076 mol) of stannic chloride to the tube containing 40 ml of methylene chloride.
4. Injected 6.5 ml (5.90 g or 0.00566 mol) of styrene monomer to the tube containing 30 ml methylene chloride. After this tube has cooled for about 10 min in the ice bath to 0°C, inject with 4 ml of the cool stannic chloride solution (cationic initiator) from the other test tube. Recorded the time it is injected (start of polymerization).
5. After 10 min, remove the tube from the ice bath and pour its contents into a flask containing 100 ml methanol to precipitate the polymer.
6. Filter and wash the polymer with methanol and then air dry in a vacuum oven for 30 min.
7. Weigh the polystyrene to the nearest 0.01 g and calculate the yield.

NOTES

1. Wash the styrene monomer twice with 25-ml portions of 25% aqueous sodium hydroxide in order to remove the inhibitor. Then wash twice with 25 ml of distilled water to remove any residual caustic.
2. Using molecular sieves, dry the styrene monomer and store immediately with a nitrogen blanket in a refrigerator.
3. Dry the test tubes in an oven and cool in a dessicator.
4. The solvents used in this procedure should be taken from freshly opened bottles and dried with 3 Å molecular sieves; they may have to be filtered if there is any suspended drying agent.
5. Use pure $SnCl_4$ as is. Where possible, use freshly prepared $SnCl_4$.
6. After using stannic chloride ($SnCl_4$), immediately clean the syringe by rinsing with dilute hydrochloric acid, water, and then methanol.

REFERENCES

1. P. H. Plesch, ed., "Cationic Polymerization and Related Complexes." Academic Press, New York, 1954.
2. P. H. Plesch, ed., "The Chemistry of Cationic Polymerization." Pergamon Press, New York, 1964.
3. F. S. Dainton and G. B. B. M. Sutherland, *J. Polym. Sci.* **4,** 37 (1949).
4. A. G. Evans, B. Holden, P. H. Plesch, M. Polanyi, H. A. Skinner, and M. A. Weinberger, *Nature (London)* **157,** 102 (1946).
5. A. G. Evans, G. W. Meadows, and M. Polanyi, *Nature (London)* **158,** 94 (1946).
6. D. C. Pepper, *Quart. Rev. Chem. Soc.* **8,** 88 (1954).
7. D. C. Pepper, in "Friedel-Crafts and Related Reactions" (G. A. Olah, ed.), Vol. II, p. 123. Wiley (Interscience), New York, 1964.
8. A. G. Evans and G. W. Meadows, *J. Polym. Sci.* **4,** 359 (1949).
9. R. G. Heiligmann, *J. Polym. Sci.* **6,** 155 (1950).
10. J. A. Bittles, A. K. Chandhuri, and S. W. Benson, *J. Polym. Sci. Part A* **2,** 1221 (1964).
11. G. F. Endres and C. G. Overberger, *J. Am. Chem. Soc.* **77,** 2201 (1955).
12. D. O. Jordan and A. R. Mathieson, *J. Chem. Soc., London* p. 611 (1952).
13. J. A. Bittles, A. K. Chandhuri, and S. W. Benson, *J. Polym. Sci. Part A* **2,** 1221 (1964).
14. F. C. Whitmore, *Ind. Eng. Chem.* **26,** 94 (1964).
15. R. Simha and L. A. Wall, in "Catalysis" (P. H. Emmett, ed.), Vol. VI, p. 266. Van Nostrand-Reinhold, Princeton, NJ, 1958.
16. J. Hine, "Physical Organic Chemistry," p. 219. McGraw-Hill, New York, 1956.
17. Y. Tsuda, *Macromol. Chem.* **36,** 102 (1960).
18. J. P. Kennedy and E. Maechal, "Carbocationic Polymerization," Wiley, New York, 1981.
19. G. A. Olah, ed., "Friedel-Crafts and Related Reactions," Vol. I. Wiley (Interscience), New York, 1963.
20. D. N. P. Satchell, *J. Chem. Soc. London* pp. 1453 and 3822 (1961).
21. A. M. Eastham, *Encycl. Polym. Sci. Technol.* **3,** 35 (1965).
22. J. P. Kennedy, *J. Org. Chem.* **35,** 532 (1970).
23. J. P. Kennedy, in "Polymer Chemistry of Synthetic Elastomers" (J. P. Kennedy and E. Tarnquist, eds.), Part 1, Chapter 5A, p. 291. Wiley (Interscience), New York, 1968; Belgian Patent 663,319, 1965.
24. J. P. Kennedy and R. M. Thomas, *Adv. Chem. Ser.* **34,** Chapt. 7 (1962).

B

POLYMERIZATION OF ACRYLIC ESTERS

INTRODUCTION

Ester polymers of methacrylic and acrylic acid are important in a wide range of applications. They are used in dental materials, glazing, adhesives, plastic bottles, elastomers, floor polishes, paint bases, plastic films, and leather finishes, to mention only a few.

For most of these esters, the free radical polymerization procedures are very similar to each other. With minor modifications, the considerations and preparations given here may be applied to many of the other common "vinyl" monomers such as styrene, vinyl acetate, vinylidene chloride, acrylonitrile, and acrylamide.

From the point of view of the organic chemist, the suspension and emulsion techniques are perhaps the best methods for preparing reasonable quantities of many homo- and copolymers. The apparatus and manipulations resemble those of familiar laboratory operations.

REACTANTS AND REACTION CONDITIONS

Inhibitors and Their Removal

As normally supplied, acrylic esters are inhibited to enhance the shelf life. Aside from dissolved oxygen, inhibitors that are deliberately added include phenolic compounds such as hydroquinone (HQ) and *p*-methoxyphenol (MEHQ, i.e., "methyl ether of hydroquinone"). These inhibitors are usually present in concentrations of 50 to 100 parts per million (ppm) by weight. Oxidation products of the phenolic inhibitors may also be present.

Inhibitors may be removed from acrylic monomers by repeated extraction of the monomer specimen with cold 0.5% aqueous sodium hydroxide solution followed by enough washes with deionized water until the last wash is substantially neutral. Then the monomer is dried over conventional drying agents, such as calcium chloride, sodium sulfate, or magnesium sulfate.

A simpler, more thorough, and more rapid method of removing inhibitors and oxidative impurities from nonacidic liquid monomers consists of passing the inhibited monomer through a short chromatography column ($\sim$ 25 cm long and 2.5 cm in diameter) packed to a height of approximately 15 cm with a coarse, dry aluminum oxide such as Alcoa CG20. (Warning: Do not use fine chromatography grades of alumina as these tend to block up rapidly and may even initiate polymerization in the column.) The effectiveness of this column treatment can be judged readily by the progress of a colored band down the column. The colored band usually stays near the top of the column and is probably caused by the inhibitor and its oxidation products.

If the inhibitor-free monomer is not used promptly, it may be stored in an appropriate refrigerator.

Free radicals should initiate polymerization efficiently. Some peroxides such as dialkyl peroxides and peresters tend to abstract hydrogen from the monomer more readily than they react to initiate polymerizations. Consequently, their efficiency as initiators is reduced.

Reaction Temperature

Other factors being equal, the higher the reaction temperature, the lower the average molecular weight of the product.

This simple, reciprocal relationship may, however, be offset by the effect of the reaction temperature on the rate of decomposition of the initiator, the number of efficiently active free radicals that form, the reactivity of the free radicals, and the effect on chain-transfer agents, if any are present.

The viscosity of the reacting system is also temperature dependent. The diffusion of the monomer and of the growing polymer chains and the heat transfer properties of the system are modified as the viscosity of the system increases or as the molecular weight of the polymer grows.

Chain-Transfer Agents

A variety of compounds may act to reduce the average molecular weight of the polymer produced by a chain-transfer mechanism during polymerization. As indicated earlier, solvents may act as chain-transfer agents, although their activity is usually low. The most commonly used agents are mercaptans, particularly the higher molecular weights ones such as dodecyl mercaptan. Naturally, such reagents may give rise to serious odor problems.

Halogenated compounds such as carbon tetrachloride and chloroform have particularly high chain-transfer constants. However, these compounds must be used with extreme caution as explosive polymerizations have been observed.

The activity of chain-transfer reagents is a function of the reaction temperature, concentration, and monomer type.

Initiators

In the polymerization of acrylic monomers by bulk, suspension, or in organic solution, the most common initiators are diacyl peroxide (e.g., dibenzoyl peroxide supplied as a paste in water) or azo compounds (e.g., 2,2'-azobisisobutyronitrile). For emulsion or aqueous solution polymerizations, sodium persulfate by itself or in combination with bisulfites or a host of other reducing agents may be used.

Whereas the literature frequently suggests the use of ammonium persulfate, this reagent is not very storage stable and consequently a sample of this reagent may not be very active. Potassium persulfate is a useful initiator, but its water solubility and rate of dissolution are not as great as those of its sodium analog. These properties may be significant when solutions of the initiator have to be added to a reaction.

Walling [1] lists four factors that should be considered in the selection of an initiator.

1. The initiator must produce free radicals at a reasonably constant rate during the polymerization process.

2. The reactive radicals have to be "available" to initiate polymerization. The homolytic decomposition of an initiator to pairs of radicals may be such that some of the radicals may recombine before they react with a monomer molecule.

3. An initiator must be stable toward induced decomposition from its own radicals or from the growing radical-terminated polymer chain in the reaction medium. If radicals induce initiator decomposition, the resultant products tend to form polymers of low average molecular weight.

4. Initiator fragments must efficiently initiate chains.

PROCEDURES

Most of the common acrylic esters may be homopolymerized by relatively simple procedures. Variations in the methods may be made because of requirements related to the final application of the polymer, limitations set by available laboratory equipment, the reactivity of the monomers, and the physical state of the monomer or of the polymer.

Bulk Polymerization

The conversion of a monomer to a polymer in the absence of diluents or dispersing agents is termed a "bulk" polymerization.

Samples of a polymer may be prepared in a test tube by simply heating the monomer with a small amount of an initiator. A handy variation of this is the test tube photopolymerization given below.

It should be noted that simple poly(methacrylates) are usually rigid and therefore either slide out of a test tube or can be isolated by breaking the test tube. Polyacrylates, however, tend to be elastomeric and frequently adhere to glass surfaces. Therefore, it is good practice to coat surfaces with "parting agents" such as a soap solution, films deposited by evaporation of poly(vinyl alcohol) solutions, silicone coatings, or fluorocarbon coatings prior to introducing the monomer. If the reaction is carried out at sufficiently low temperatures, polyethylene or Teflon equipment may be used.

Several other factors must be kept in mind, particularly in bulk and suspension polymerizations.

1. Polymerizations of acrylic and methacrylic esters are highly exothermic (e.g., $\Delta H_{polymerization}$ of ethyl acrylate is 13.8 kcal/mol [2]). Generally, the heats of polymerization of acrylates are greater than those of methacrylates.

2. Frequently, even if as little as 20% of the monomer has polymerized, an autoaccelerating polymerization effect will take place. This may manifest itself in an increase in the heat evolved as the process nears completion. Particularly in large-scale, industrial polymerizations, this effect, known as the "Trommsdorff effect" or "gel effect," may be quite dangerous. In fact, serious explosions have

been attributed to it [3–13]. The effect is associated with a rapid increase in the average molecular weight of the polymer. It is assumed that as polymerization progresses, the termination step of the chain process is prevented because of the increasing viscosity of the system. The increased viscosity also reduces the heat transfer rate of the system.

3. Because the density of a polymer is substantially higher than that of the corresponding monomer, there is considerable shrinkage of the volume of the material. In the case of methyl methacrylate, this shrinkage, at 25°C, amounts of 20.6–21.2% [14].

The Percentage shrinkage is readily estimated by

$$\% \text{ shrinkage} = 100(D_p - D_m)/D_p,$$

where D_m is the density of the monomer at 25°C and D_p is the density of the polymer at 25°C.

4. In most cases, a small amount of unreacted monomer remains in the polymer. Frequently, this residual monomer may be converted by a posttreatment of the polymer at elevated temperatures or by exhaustive warming under reduced pressure [3–13,15,16].

Suspension Polymerization

A sharp distinction must be drawn between suspension (or slurry) and emulsion polymerization processes.

The term *suspension polymerization* refers to the polymerization of macroscopic droplets in an aqueous medium. The kinetics is essentially that of a bulk polymerization with the expected adjustments associated with carrying out a number of bulk polymerizations in small particles more or less simultaneously and in reasonably good contact with a heat exchanger (i.e., the reaction medium) to control the exothermic nature of the process. Usually, suspension polymerizations are characterized by the use of monomer-soluble initiators and the use of suspending agents.

However, *emulsion polymerizations* involve the formation of colloidal polymer particles that are essentially permanently suspended in the reaction medium. The reaction mechanism involves the migration of monomer molecules from liquid monomer droplets to sites of polymerization that originate in micelles consisting of surface-active agent molecules surrounding monomer molecules. Emulsion polymerizations are usually characterized by the requirement of surfactants during the initiation of the process and by the use of water-soluble initiators. This process also permits good control of the exothermic nature of the polymerization.

Polymerizations that are carried out in nonaqueous continuous phases instead of water are termed *dispersion polymerizations* regardless of whether the product consists of filterable particles or of a nonaqueous colloidal system.

Suspension polymerizations are among the most convenient laboratory procedures as well as plant procedures for the preparation of polymers. The advantages of this method include wide applicability (it may be used with most water-insoluble or partially water-soluble monomers), rapid reaction, ease of temperature control, ease of preparing copolymers, ease of handling the final product, and control of particle size.

In this procedure, the polymer is normally isolated as fine spheres. The particle size is determined by the reaction temperature, the ratio of monomer to water, the rate and efficiency of agitation, the nature of the suspending agent, the suspending agent concentration, and, of course, the nature of the monomer. With increasing levels of suspending agent, the particle size decreases.

It is a good policy, when first experimenting with a given system, to have a measured quantity of additional suspending agent ready at hand. Then, if incipient agglomeration of particles is observed, additional suspending agent can be added rapidly. In subsequent preparations, this additional quantity of suspending agent may be added from the start. If excess suspending agent is used, emulsification of the monomer may take place and a polymer latex may be produced along with polymer beads.

Common suspending agents are poly(vinyl alcohols) of various molecular weights and degrees of hydrolysis, starches, gelatin, calcium phosphate (especially freshly precipitated calcium phosphate dispersed in water to be used in the preparation), salts of poly(acrylic acid), gum arabic, gum tragacanth, etc.

Initiators commonly used include dibenzoyl peroxide, lauryl peroxide, 2,2′-azobis isobutyronitrile, and others that are suitable for use in the temperature range of approximately 60–90°C.

The hazard of agglomeration is greatest when acrylates are polymerized. The products tend to be elastomers and, in the course of the polymerization of these monomers, they tend to go through a "sticky stage." However, the proper selection of the suspending agent frequently prevents agglomeration.

The suspension process may be carried out not only with compositions consisting of a solution of the initiator in the monomer, but also with complex mixtures that incorporate plasticizers, pigment particles, chain-transfer agents, and modifiers, and, of course, with various comonomers.

Emulsion Polymerization

The section on suspension polymerization indicated the differentiation between suspension and emulsion (or latex) polymerizations. Emulsion polymers usually are formed with the initiator in the aqueous phase, in the presence of surfactants, and with polymer particles of colloidal dimensions, i.e., on the order of 0.1 μm in diameter [17]. Generally, the molecular weights of the polymers produced by an emulsion process are substantially greater than those produced by bulk or suspension polymerizations. The rate of polymer production is also higher. As a large quantity of water is usually present, temperature control is often simple.

Typical emulsion polymerization recipes involve a large variety of ingredients. Therefore, the possibilities of variations are many. Among the variables to be considered are the nature of the monomer or monomers, the nature and concentration of surfactants, the nature of the initiating system, protective colloids and other stabilizing systems, cosolvents, chain-tranfer agents, buffer systems, "short stops," and other additives for the modification of latex properties to achieve the desired end properties of the product.

The ratio of total nonvolatiles to water (usually referred to as "percentage solids") is also important. When starting experimental work in emulsion polymerization it is best to develop the techniques required to prepare 35–40% solid latices without the formation of coagula. Latices with higher solid content are more difficult to prepare. The geometry of close packing of uniform spheres imposes a limit on the percentage nonvolatiles at approximately 60–65%. Dissolved nonvolatile components and the judicious packing of spheres of several diameters may permit the formation of more concentrated latexes, in principle.

In the preparation of a polymer latex, the initial relationship of water, surfactant, and monomer concentration determines the number of particles present in the reaction vessel. Once the process is underway, further addition of monomer does not change the number of latex particles. If such additional monomer polymerizes, the additional polymer is formed on the existing particles. As expected, smaller initial particles imbibe more of the additional monomer than larger ones. Consequently, a procedure in which monomer is added to preformed latex polymer tends to produce a latex with a uniform particle size, i.e., a "monodispersed latex." As the stability of the latex is dependent to a major extent on the effective amount of surfactant on a particle surface, a considerable increase of the volume of the latex particles is possible with minor increases of the surface area purely on geometric grounds (an increase of the volume of a sphere by a factor of 8 increases the surface area by a factor of 4, whereas the particle diameter only doubles). These considerations have many practical applications, not the least of which is the possibility of preparing latex particles started with one comonomer composition to which a different comonomer solution is added.

From the preparative standpoint, there are two classes of initiating systems.

1. The thermal initiator system. This system is made up of water-soluble materials that produce free radicals at a certain temperature to initiate polymerization. The most commonly used materials for such thermal emulsion polymerizations are potassium persulfate, sodium persulfate, or ammonium persulfate.

2. Activated or redox initiation systems. Because these systems depend on the generation of free radicals by the oxidation–reduction reactions of water-soluble compounds, initiation near room temperature is possible. In fact, redox systems operating below room temperature are available (some consist of organic hydroperoxides dispersed in the monomer and a water-soluble reducing agent). A typical redox system consists of sodium persulfate and sodium metabisulfite. There is some evidence, particularly in the case of redox polymerizations, that traces of iron salts catalyze the generation of free radicals. Frequently these iron salts are supplied by impurities in the surfactant (quite common in the case of surfactants specifically manufactured for emulsion polymerization) or by stainless-steel stirrers used in the apparatus. In other recipes, iron salts may be supplied in the form of ferrous ammonium sulfate or, if the pH is low enough, in the form of ferric salts.

In particular, if a latex is to be used for coatings, adhesives, or film applications, no silicone-base stopcock greases should be used on emulsion polymerization equipment. Although hydrocarbon greases are not completely satisfactory either, there are very few alternatives. Teflon tapes, sleeves, and stoppers may be useful, although expensive.

REFERENCES

1. C. Walling, *Polym. Prep. Am. Chem. Soc. Div. Polym. Chem.* **11**(2), 721 (1970).
2. L. S. Luskin and R. J. Meyers, *Encycl. Polym. Sci. Technol.* **1**, 246 (1964).
3. E. Trommsdorff, H. Köhle, and P. Lagally, *Makromol. Chem.* **1**, 169 (1948).
4. M. S. Matheson, E. E. Auer, E. B. Bevilacqua, and E. J. Hart, *J. Am. Chem. Soc.* **71**, 497 (1949).
5. G. Odian, M. Sobel, A. Rossi, and R. Klein, *J. Polym. Sci.* **55**, 663 (1961).
6. V. E. Shashoua and K. E. Van Holde, *J. Polym. Sci.* **28**, 395 (1958).
7. A. T. Guertin, *J. Polym. Sci. Part B* **1**, 477 (1963).
8. K. Horie, I. Mita, and H. Kambe, *J. Polym. Sci. Part A-1* **6**, 2663 (1968).
9. G. Henrici-Olivé and S. Olivé, *Makromol. Chem.* **27**, 166 (1958).
10. Kunstst.-Plast. (Solothurn) **5**, 315 (1958).
11. G. V. Schulz, *Z. Phys. Chem.* **8**, 290 (1956).
12. M. Gordon and B. M. Grieveson, *J. Polym. Sci.* **17**, 107 (1955).

13. G. V. Korolev *et al., Vysokomol. Soedin.* **4**(10), 1520 (11), 1654 (1962).
14. E. H. Riddle, "Monomeric Acrylic Esters." Van Nostrand-Reinhold, Princeton, NJ, 1954.
15. T. M. Laakso and C. C. Unruh, *Ind. Eng. Chem.* **50,** 1119 (1958).
16. R. H. Wiley and G. M. Brauer, *J. Polym. Sci.* **3,** 647 (1948).
17. F. W. Billmeyer, Jr., "Textbook of Polymer Science." 2nd Ed., Wiley (Interscience). New York, 1971.

EXPERIMENT 5

Bulk Photopolymerization of Methyl Methacrylate: A Test Tube Demonstration

INTRODUCTION

The initiation of polymerization by ultraviolet radiation has been of particular interest in the study of free radical processes [1,2]. The test tube demonstration described here is simple and may be used to evaluate the polymerizability of new monomers or to study some of the physical properties of a polymer. Although the method is particularly effective for acrylic and methacrylic monomers, it may also be applied to the polymerization of a wide range of "vinyl"-type monomers.

Generally, the method depends on the sensitization of the monomer to ultraviolet radiation with reagents such as biacetyl or benzoin [2,3].

With sensitizers, initiation stops when the source of radiation is turned off, which is followed by a rapid decay of the polymerization process. When a conventional initiator, such as dibenzoyl peroxide, is also present, the process is more rapid than when the sensitizer is used by itself. It also seems to continue after the radiation source has been discontinued. It is presumed that ultraviolet (UV)-induced decomposition of the peroxide becomes involved in the process. By this method, polymerizations may be carried out at temperatures well below those normally used with thermal initiators such as organic peroxides.

SAFETY PRECAUTIONS

Before this experiment is carried out, the student must read the material safety data sheets (MSDS) for all the chemicals used as well as for the products. The instructor must approve that you have read and understood the MSDS for the safe handling of these materials.

Please be advised that all chemicals should be considered hazardous and should be handled in a hood and with proper personal protective equipment (lab coat, proper gloves, approved safety glasses, and/or goggles). Avoid inhaling vapors and/or aerosolized materials. Avoid skin/eye contact with all chemicals at all times. Wash hands frequently. See the instructor if you have any questions or concerns.

APPARATUS

1. A No. 9800 Pyrex-brand test tube, 12.5 cm long, i.d. ~12 mm
2. A No. 0 neoprene stopper
3. Polyethylene film, ~5 × 5 cm
4. Balance, weights, weighing paper, or dishes
5. Graduated cylinder, 25-ml capacity
6. Test tube rack (or tall form beaker)
7. Miscellaneous laboratory equipment such as beakers, paper towels, aluminum foil, and stirring rods
8. Laboratory hood
9. Space on a sunny window sill or some other secure, sunny location

REAGENTS AND MATERIALS

1. 20 ml of a 0.5% solution of sodium stearate (soap) in isopropanol
2. 10 ml of methyl methacrylate (commercial grade)
3. 0.5 g of benzoin
4. 0.5 g of dibenzoyl peroxide

PROCEDURE

1. Fill the 12 cm × 12 mm test tube with some of the 0.5% solution of sodium stearate in isopropanol. In a hood, pour the solution off into a suitable container. Allow the inside of the test tube to dry thoroughly (see Note 1).
2. While the test tube dries, prepare a solution of 0.5 g of benzoin in 10 ml of methyl methacrylate. Then add 0.5 g of dibenzoyl peroxide (see Note 2).
3. When the test tube is ready for use, place the methyl methacrylate–initiator solution in the test tube. Cover the neoprene stopper with the polyethylene film and stopper the test tube.
4. Place the test tube in a test tube rack or a beaker. Note the time and expose the test tube to sunlight. From time to time, note the time and observe the progress of the polymerization (see Note 3).

NOTES

1. In this experiment, coat the inside of the test tube with a thin layer of sodium stearate (or a plain, nonperfumed soap). This will assist in removing the product from the apparatus.
2. In the laboratory, the source of radiation may be a high-intensity mercury lamp. However, for simple test purposes, sunlight is quite suitable. Even a lightly overcast sky furnishes sufficient UV radiation for photo-induced polymerizations. Naturally, bright sunlight is more effective.
3. As the polymerization of methyl methacrylate is somewhat inhibited by atmospheric oxygen, the upper layer of the polymerizing monomer may still be

fluid, even though most of the material has already been converted. Therefore, when checking on the progress of the experiment, do not simply shake the test tube. Instead, either invert the test tube or insert a probe (a long pin, a paper clip, etc.) to check on the process.

REFERENCES

1. P. J. Flory, "Principles of Polymer Chemistry," p. 149ff. Cornell Univ. Press, Ithaca, NY, 1953.
2. C. E. Schildknecht, "Vinyl and Related Polymers," p. 207ff. Wiley, New York, 1952.
3. W. E. F. Gates, British Patent 566,795, 1945.

EXPERIMENT 6

Suspension Polymerization of Methyl Methacrylate

INTRODUCTION

Details of this experiment may be found in Refs. 1 and 2.

SAFETY PRECAUTIONS

Before this experiment is carried out, the student must read the material safety data sheets (MSDS) for all the chemicals used as well as for the products. The instructor must approve that you have read and understood the MSDS for the safe handling of these materials.

Please be advised that all chemicals should be considered hazardous and should be handled in a hood and with proper personal protective equipment (lab coat, proper gloves, approved safety glasses, and/or goggles). Avoid inhaling vapors and/or aerosolized materials. Avoid skin/eye contact with all chemicals at all times. Wash hands frequently. See the instructor if you have any questions or concerns.

APPARATUS

1. 500-ml three-necked flask
2. Reflux condenser
3. Sealed stirrer with stainless-steel stirring rod and appropriate stirring motor (see Note 2)
4. Thermometer
5. Calibrated addition funnel
6. Water bath arranged so that it may be raised or lowered as needed
7. Balance, weights, weighing paper, or dishes
8. 100-ml graduated cylinder
9. Cheesecloth or nylon chiffon (possibly a nylon stocking)
10. Miscellaneous laboratory equipment
11. Laboratory hood

12. A large glass funnel
13. A 2-liter beaker with a few boiling chips
14. Hot plate
15. Vacuum oven

REAGENTS AND MATERIALS

1. 150 ml of a 1% solution of sodium poly(methacrylate) in water
2. 50 ml of a 5% solution of sodium poly(methacrylate) in water (see Note 1)
3. A buffer solution of 0.85 g of disodium phosphate and 0.05 g of monosodium phosphate in 5 ml of water
4. 0.5 g of dibenzoyl peroxide paste in water
5. 50.0 g of methyl methacrylate (inhibitor free)
6. Deionized water (several liters)

PROCEDURE

1. In a laboratory hood, equip a 500-ml three-necked flask with a sealed stirrer and motor, condenser, thermometer, and an addition funnel. Add 150 ml of a 1% solution of sodium poly(methacrylate) in water and a buffer solution of 0.85 g of disodium phosphate and 0.05 g of monosodium phosphate in 5 ml of water.
2. Add a dispersion of 0.5 g of dibenzoyl peroxide in 50 g of inhibitor-free methyl methacrylate to the reaction flask.
3. Measure 25 ml of the 5% aqueous solution of sodium poly(methacrylate) into the addition funnel.
4. Place the assembled reaction flask in the water bath, attach the stirring motor, and begin stirring.
5. Adjust the stirrer speed so that droplets of monomer form that are 2–3 mm in diameter.
6. Heat the flask with the water bath at 80–82°C for 45 to 60 min (see Notes 3 and 4).
7. Collect the solid particles on a funnel fitted with cheesecloth.
8. Bundle the particles in the cheesecloth, and place the bundle in 1 liter of boiling deionized water. Remove the bundle and wash two more times in fresh boiling deionized water.
9. Open the bundle. Dry the product under reduced pressure at 60–70°C.

NOTES

1. Additional amounts of the suspended agent solution should only be used to counteract agglomeration of particles that may occasionally occur.
2. Sealed stirrers may be prepared by drilling a poly(ethylene) stopper (e.g., Aldrich Z 10.579-1, which fits a standard taper joint) to hold a 10-mm stainless-steel stirrer rod snugly. Other possibilities are a Safe-Lab stirrer bearing or simply a stopper of the appropriate size drilled to hold a glass tube with an inside diameter slightly larger than the stirring rod that is to be used. A short piece of neoprene tubing may be used to seal the top of the tube to the rod. This sealing tubing must just barely touch the rod. If necessary, a small amount of petroleum jelly may be used to lubricate the seal. Under no circumstances should silicone grease be used in these preparations.

3. The bath may have to be raised or lowered to maintain the reaction temperature after an initial heat evolution.

4. If a tendency for agglomeration of the particles is observed during the process, measured amounts of the 5% suspending agent solution can be added to control this. The actual volume used should be recorded. In future experiments, the suspending agent requirement may be adjusted in light of this experience.

REFERENCES

1. H. P. Wiley, *J. Chem. Ed.* **25,** 204 (1948).
2. D. P. Hart, *Macromol. Syn.* **1,** 22 (1963).

EXPERIMENT 7

Redox Emulsion Polymerization of Ethyl Acrylate

INTRODUCTION

Details of this experiment may be found in Ref. 1.

SAFETY PRECAUTIONS

Before this experiment is carried out, the student must read the material safety data sheets (MSDS) for all the chemicals used as well as for the products. The instructor must approve that you have read and understood the MSDS for the safe handling of these materials.

Please be advised that all chemicals should be considered hazardous and should be handled in a hood and with proper personal protective equipment (lab coat, proper gloves, approved safety glasses, and/or goggles). Avoid inhaling vapors and/or aerosolized materials. Avoid skin/eye contact with all chemicals at all times. Wash hands frequently. See the instructor if you have any questions or concerns.

APPARATUS

1. Laboratory hood
2. Water bath
3. 1-liter resin kettle equipped with sealed stirrer and appropriate stirring motor, reflux condenser, Claissen adapter, thermometer, nitrogen inlet tube, which may be raised or lowered in the kettle, and glass topper (a 1-liter four-necked flask with a Claissen adaptor may be used instead of the resin kettle)
4. Appropriate rings, ring stand or rack, and clamps
5. A supply of nitrogen with required pressure-reducing valves
6. Balance, weights, graduated cylinders, beakers, etc.
7. Cheesecloth or nylon chiffon
8. 50-ml disposable aluminum dishes (e.g., Aldrich, Z 15.485-7), tared

9. Vacuum oven
10. Large and small glass funnels
11. 2-liter beakers
12. Miscellaneous laboratory equipment
13. A straight glass rod, ~20 cm long
14. Several lengths of 0.5-in. transparent adhesive tape
15. 10-ml pipette
16. A flat glass plate, ~30 × 30 cm

REAGENTS AND MATERIALS

1. 376 ml of deionized water
2. 2 g of sodium lauryl sulfate (or 6 ml of a commercial 30% solution of sodium lauryl sulfate)
3. Nitrogen (low in residual oxygen)
4. 200 g of ethyl acrylate (inhibitor free)
5. 4 ml of a solution freshly prepared from 0.3 g of ferrous sulfate heptahydrate dissolved in 200 ml of deionized water
6. 1 g sodium persulfate
7. 1 g of sodium metabisulfite
8. 5 drops of 70% commercial *tert*-butyl hydroperoxide
9. Ice
10. Hydroquinone, a few crystals

PROCEDURE

1. Place 376 ml of deionized water into a 1-liter resin kettle equipped with a sealed stirrer and appropriate stirring motor (see also Note 2 of Experiment 2), reflux condenser, thermometer, and nitrogen inlet tube reaching down to just above the stirrer blade.
2. Pass a gentle stream of nitrogen into the water for 15 min while stirring (see Note 1).
3. While stirring, add 200 g of inhibitor-free ethyl acrylate, 4 ml of the ferrous sulfate solution, and 1 g of sodium persulfate through the open neck of the kettle.
4. Raise the nitrogen inlet tube above the liquid level and reduce the gas flow rate.
5. Cool the reaction mixture with the water bath to 20°C and add 1 g of sodium metabisulfite followed by 5 drops of 70% *tert*-butyl hydroperoxide. Stopper the remaining open neck of the flask.
6. The polymerization starts rapidly (see Note 2).
7. Unless controlled by adding ice to the water bath, the temperature in the reactor may rise to approximately 90°C (see Note 3).
8. After the heat evolution has subsided, heat the product (latex) to about 90°C for 0.5 hr.
9. Cool and pass the latex through a strainer formed from a large funnel and the nylon chiffon.
10. To demonstrate film-forming properties of the latex, prepare a "drawdown" bar by adhering a double thickness of adhesive tape to each end of a glass rod.
11. Place this rod on a glass plate. Pour a small quantity of the cooled latex on one side of the rod and spread the latex by rolling the rod

smoothly in one direction across the plate. Once the liquid film has formed, remove the rod and allow the film to dry at room temperature. Note properties such as transparency, flexibility, and adhesion to glass.
12. Repeat this experiment with the latex cooled in ice and the plate at ice temperature. Allow the drying process to take place near the ice temperature. Again note the properties.

NOTES

1. A stirring rate of about 200 rpm should be adequate to keep the monomer dispersed without the formation of a separate monomer layer. Once polymerization is underway, the rate of agitation may not be important as long as no separate monomer layer forms. Once sodium lauryl sulfate has been added to the water, the nitrogen inlet tube must be raised above the liquid level to reduce foam and bubble formation.

2. Toward the start of the emulsion polymerization, the reacting mixture often changes to a milky appearance with a sky blue edge to the outer surface.

3. The progress, as well as the completion, of the process is determined by periodically withdrawing 10-ml samples from the reactor (at times that are recorded), placing the sample into a tared aluminum dish, adding a few crystals of hydroquinone, and drying the sample rapidly in a vacuum oven at 80°C to constant weight. When the dry weight in the dish corresponds to the calculated percentage solids, the process has been completed.

REFERENCE

1. "Emulsion Polymerization of Acrylic Monomers," Bull. CM-104 A/cf Rohm and Haas Company, Philadelphia, PA.

C

POLYAMIDES

The synthetic chemistry of polyamides is briefly reviewed and illustrated with a laboratory experiment showing the interfacial polycondensation method.

INTRODUCTION

Polymeric amides are found in nature in the many polypeptides (proteins) that constitute a variety of animal organisms and the composition of silk and wool. It was a search to prepare substitutes for the latter that led to the commercial development of the synthetic polyamides known as nylons.

Several early investigators reported polyamide type materials. Balbiano and Trasciatti [1] heated a mixture of glycine in glycerol to give a yellow amorphous glycine polymer [2]. Manasse [3] obtained nylon 7 by heating 7-aminoheptanoic acid to the melting point. Curtius [4] found that ethyl glycinate

$$n\text{NH}_2(\text{CH}_2)_6\text{COOH} \longrightarrow [-\text{NH}(\text{CH}_2)_6\text{CO}-]_n + n\text{H}_2\text{O} \qquad (1)$$

polymerizes on standing in the presence of moisture or in the dry state in ethyl ether to afford polyglycine [5]:

$$n\text{NH}_2\text{CH}_2\text{COOC}_2\text{H}_5 \longrightarrow [-\text{NHCH}_2\text{CO}-]_n + n\text{C}_2\text{H}_5\text{OH} \qquad (2)$$

The β amino acids on heating afford ammonia but no polymers. In addition, the γ and δ amino acids on heating afford stable lactams and no polymers. However, the ϵ, ζ, and η amino acids give polyamides.

The reaction of dibasic acids with diamines was reported in the early literature [6–13] to give low molecular weight cyclic amides as infusible and insoluble products. It was Carothers [14–18] who first recognized that polymeric amides were formed by the reaction of diamines with dibasic acids. Many of the polyamides were able to be spun into fibers and the fibers were called "Nylon" by du Pont [14–20].

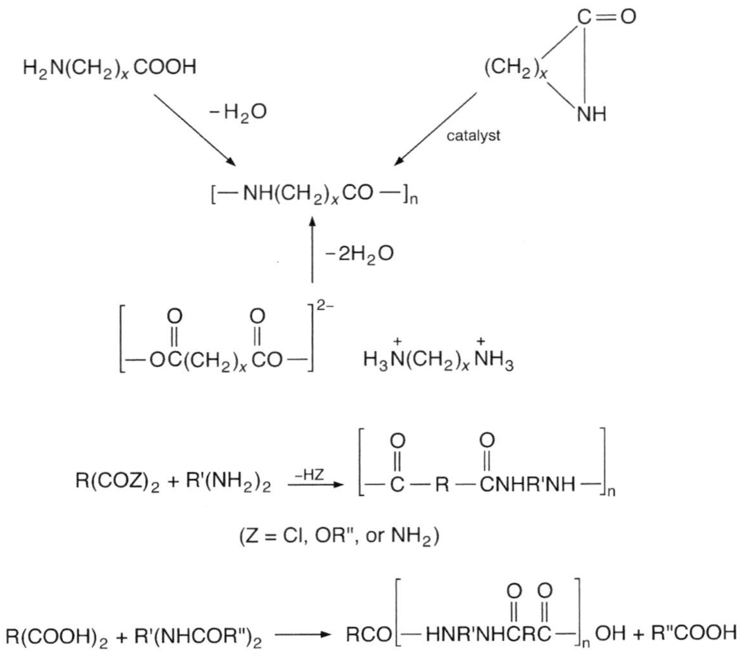

$$H_2N(CH_2)_x COOH \xrightarrow{-H_2O} [-NH(CH_2)_x CO-]_n \xleftarrow[catalyst]{} \underset{NH}{\overset{C=O}{(CH_2)_x}}$$

$$\left[-OC(CH_2)_x CO-\right]^{2-} \; H_3\overset{+}{N}(CH_2)_x \overset{+}{N}H_3 \xrightarrow{-2H_2O} [-NH(CH_2)_x CO-]_n$$

$$R(COZ)_2 + R'(NH_2)_2 \xrightarrow{-HZ} \left[-\overset{O}{\underset{\parallel}{C}}-R-\overset{O}{\underset{\parallel}{C}}NHR'NH-\right]_n$$

$$(Z = Cl, OR'', \text{ or } NH_2)$$

$$R(COOH)_2 + R'(NHCOR'')_2 \longrightarrow RCO\left[-HNR'NH\overset{O}{\underset{\parallel}{C}}R\overset{O}{\underset{\parallel}{C}}-\right]_n OH + R''COOH$$

Scheme 1 Major preparative methods for the synthesis of polyamides.

The major methods of preparing polyamides are summarized in Scheme 1. Other methods of less importance are summarized in Scheme 2.

Additional developments in polyamide synthesis involve the preparation of nylon-1 by the anionic polymerization of alkylisocyanates at temperatures below $-20°C$ [21].

$$RNCO \longrightarrow \left(-\underset{\underset{O}{\parallel}}{C}-\underset{\underset{R}{|}}{N}-\right)_n \qquad (2a)$$

Nylon 4,6 or poly(tetramethylenediamine-co-adipic acid) melts approximately 30°C higher than nylon 6,6 (i.e., 295°C vs 265°C). It is possible to prepare

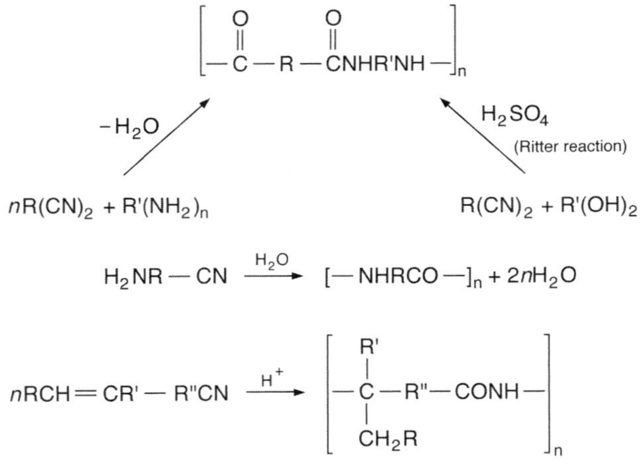

$$nR(CN)_2 + R'(NH_2)_n \xrightarrow{-H_2O} \left[-\overset{O}{\underset{\parallel}{C}}-R-\overset{O}{\underset{\parallel}{C}}NHR'NH-\right]_n \xleftarrow[\text{(Ritter reaction)}]{H_2SO_4} R(CN)_2 + R'(OH)_2$$

$$H_2NR-CN \xrightarrow{H_2O} [-NHRCO-]_n + 2nH_2O$$

$$nRCH=CR'-R''CN \xrightarrow{H^+} \left[-\underset{\underset{CH_2R}{|}}{\overset{\overset{R'}{|}}{C}}-R''-CONH-\right]_n$$

Scheme 2 Miscellaneous methods for the synthesis of polyamides from nitriles.

it by a solid-phase polymerization to give polymers with $\overline{M}_n$ of about 33,000 [22, 23]. The fibers can be spun at 305–330°C to give filaments of high modulus that are used in tire cords [24, 25].

The first commercial aramid fiber was based on poly(*m*-phenylene isophthalamide) and has the trade name Nomex (Du Pont) [26–29]:

(2b)

Du Pont also introduced Kevlar based on poly(*p*-phenylene terephthalamide) [30]:

Aramid fibers have high tensile strength and good flame resistance.

The synthesis of optically active polyamides, or nylons, is a growing area of interest. From 1980 to 1991 there have been many citations in *Chemical Abstracts* on this subject. For example, optically active polyamides have been prepared for the resolution of optical isomers. The polyamides are prepared from optically active amines or dicarboxylic acids. One polyamide was prepared from (-)-*trans*-1,2-diaminocyclohexane and terephthaloyl chloride and was used to resolve 2,2'-dihydroxy-6,6'-dimethylbiphenyl [31]. These optically active polyamides can be used in chromatography applications to resolve other optically active compositions.

In addition, optically active lactones can be polymerized to polyamides with optical activity [32].

Polyamides with molecular weights above 7000 are useful as they possess properties that allow them to be spun into fibers [33].

Cross-linked polyamides may be prepared from polyfunctional amino acids, triamines, or tricarboxylic acids.

The preparation of polyamides has been reviewed, and these reviews are worth consulting [34–37].

Some of the main uses [33,34] of polyamides or nylons are for synthetic fibers for the tire, carpet, stocking, and upholstery industries. Use of polyamides as molding and extrusion resins for the plastics industry is also of increasing importance [38].

REFERENCES

1. L. Balbiano and D. Trasciatti, *Ber. Deut. Chem. Ges.* **33**, 2323 (1900).
2. L. C. Maillard, *Ann. Chim. Anal. Chim. Appl.* **1**(9), 519 (1914); **2**(9), 210 (1914).

3. A. Manasse, *Ber. Deut. Chem. Ges.* **35,** 1367 (1902).
4. I. Curtius, *Ber. Deut. Chem. Ges.* **37,** 1284 (1904).
5. M. Frankel and A. Katchalsky, *Nature* **144,** 330 (1939).
6. E. Fischer and H. Koch, *Justus Liebigs Ann. Chem.* **232,** 227 (1886).
7. A. W. Hofmann, *Ber. Deut. Chem. Ges.* **5,** 247 (1872).
8. M. Freund, *Ber. Deut. Chem. Ges.* **17,** 137 (1884).
9. F. Anderlini, *Gazz. Chim. Ital.* **24**(1) 397 (1894).
10. E. Fischer, *Ber. Deut. Chem. Ges.* **46,** 2504 (1913).
11. H. Meyer, *Justus Liebigs Ann. Chem.* **347,** 17 (1906).
12. P. Ruggli, *Justus Liebigs Ann. Chem.* **392,** 92 (1912).
13. C. L. Butler and R. Adams, *J. Am. Chem. Soc.* **47,** 2614 (1925).
14. W. H. Carothers and G. J. Berchet, *J. Am. Chem. Soc.* **52,** 5289 (1930).
15. W. H. Carothers and J. W. Hill, *J. Am. Chem. Soc.* **54,** 1566 (1932).
16. W. H. Carothers, U.S. Patent 2,071,250, 1937.
17. W. H. Carothers, U.S. Patent 2,130,523, 1938.
18. W. H. Carothers, U.S. Patent, 2,130,947, 1938.
19. W. H. Carothers, U.S. Patent 2,130,948, 1938.
20. H. K. Livingston, M. S. Sioshansi, and M. D. Glick, *Macromol. Sci. Rev. Macromol. Chem.* **6**(1), **29** (1971).
21. V. L. Shashoua, W. W. Sweeny, and R. F. Tietz, *J. Am. Chem. Soc.* **82,** 866 (1960).
22. R. J. Gaymans and E. H. J. P. Bour, Neth. Patent Applic. 80 01,763 and 80 01,764 (10/16/81).
23. E. H. J. P. Bour and J. M. M. Warnier, Eur. Patent Applic. 77, 106 (4/20/83).
24. Japanese Patent Ko Kai-Tokkyo Kho 59 88,910 (5/23/84).
25. Japanese Patent Kohai Tokkyo Kho 59 76,914 (5/2/84).
26. J. Preston, *Encycl. Polym. Sci. Eng.* **11,** 381 (1988).
27. J. W. Hannell, *Polym. News* **1**(1), 8 (1925).
28. C. W. Stephens, U.S. Patent 3,049,518 (8/14/62).
29. F. W. King, U.S. Patent 3,079,219 (2/26/63).
30. R. E. Wilfong and J. Zimmerman. *J. Applied Polym. Sci.* **17,** 2039 (1973).
31. Y. Okamoto and K. Hatada, Japan Pat. Jpn. Kokai Tokkyo Koho JP 01014237A2 (1/18/89); *Chem. Abstr.* **111,** 154615b (1989).
32. F. Carriere, C. Bui, R. Blottiau, and H. Sekiguchi, in "Proc. IUPAC. Macromol. Symp. 28th Oxford," p. 126, UK, (1982), **99,** 123052v (1983).
33. O. E. Snider and R. J. Richardson, *Encycl. Polym. Sci. Technol.* **10,** 347 (1969).
34. W. Sweeney and J. Zimmerman, *Encycl. Polym. Sci. Technol.* **10,** 483 (1961).
35. J. Zimmerman, *Encycl. Polym. Sci. Eng.* **11,** 315 (1988).
36. J. Preston, *Encycl. Polym. Sci. Eng.* **11,** 381 (1988).
37. R. J. Welgos, *Encycl. Polym. Sci. Eng.* **11,** 410 (1988).
38. E. C. Schule, *Encycl. Polym. Sci. Technol.* **10,** 460 (1969).

EXPERIMENT 8

Preparation of Poly(hexamethylenesebacamide) (Nylon 6-10) by an Interfacial Polymerization Technique

$$(CH_2)_6(NH_2)_2 + (CH_2)_8(COCl)_2 \rightarrow \left[-HN-(CH_2)_6-NH\overset{O}{\underset{\|}{C}}-(CH_2)_8-\overset{O}{\underset{\|}{C}}- \right]_n$$

INTRODUCTION

Details of this experiment may be found in Ref. 1. The interfacial polymerization method to prepare polyamides involves the reaction of a diacid dichloride with a diamine between two immiscible liquids as the reaction zone (with or without stirring). The method is useful where the reactants are sensitive to high temperature and where the polymer degrades before the melt point is reached (as in melt polymerization techniques).

BACKGROUND

The use of the low-temperature interfacial condensation technique to prepare polyamides and various other polymers has been reviewed in Refs. 2 and 3.

Some of the important variables involved in interfacial polymerization are the (a) organic solvent, (b) reactant concentration, and (c) use of added detergents [4].

SAFETY PRECAUTIONS

Before this experiment is carried out, the student must read the material safety data sheets (MSDS) for all the chemicals used as well as for the products. The instructor must approve that you have read and understood the MSDS for the safe handling of these materials.

Please be advised that all chemicals should be considered hazardous and should be handled in a hood and with proper personal protective equipment (lab coat, proper gloves, approved safety glasses, and/or goggles). Avoid inhaling vapors and/or aerosolized materials. Avoid skin/eye contact with all chemicals at all times. Wash hands frequently. See the instructor if you have any questions or concerns.

APPARATUS

1. Tall-form beaker
2. Graduated cylinder
3. Tweezers
4. Balance

REAGENTS AND MATERIALS

1. Sebacoyl chloride (reagent grade or freshly distilled)
2. Tetrachloroethylene (reagent grade or freshly distilled)
3. Hexamethylenediamine (reagent grade or freshly distilled)
4. 50% aqueous ethanol

PROCEDURE

1. Add a solution of 3.0 ml (0.014 mol) of sebacoyl chloride dissolved in 100 ml of anhydrous tetrachloroethylene as received (see Note 1) to a tall-form beaker.
2. Carefully pour a solution of 4.4 g (0.038 mol) of hexamethylenediamine (see Note 2) dissolved in 50 ml of water over this acid chloride solution.
3. Grasp the polyamide film that begins to form at the interface of these two solutions with tweezers or a glass rod and slowly pull it out of the beaker in a continuous fashion. Stop the process when one of the reactants becomes depleted.
4. Wash the resulting "rope"-like polymer with 50% aqueous ethanol or acetone, dry, and weigh to afford 3.16–3.56 g (80–90%) yields of polyamide, η_{inh} = 0.4–1.8 (m-cresol, 0.5% conc. at 25°C), m.p. 215°C (soluble in formic acid (see Note 3).

NOTES

1. The organic solvent is the most important variable as it controls partition and diffusion of the reactants between the two immiscible phases, the reaction rate, solubility, and swelling of permeability of the growing polymer. The solvent should be of such composition so as to prevent precipitation of the polymer before a high molecular weight has been attained. The final polymer should not dissolve in the solvent. The type of solvent will influence the characteristics of the physical state of the final polymer. Solvents such as chlorinated or aromatic hydrocarbons make useful solvents in this system.

Concentrations in the range of approximately 5% polymer based on the combined weights of water and organic solvent usually are optimum. Concentrations too low may lead to hydrolysis of the acid halide and concentrations too high may cause excessive swelling of the solvent in the polymer.

In some cases the addition of 0.2–1% of sodium lauryl sulfate has been found to give satisfactory results. In many cases it may not be necessary.

The reactants should be pure but need not be distilled prior to use. A slight excess (5–10%) of diamine usually helps produce higher molecular weights.

The advantage of the interfacial polymerization process is that it is a low-temperature process requiring ordinary equipment. It also allows one to prepare those polyamides that are unstable in the melt polymerization process. Random or block polymers can be prepared easily depending on the reactivity of the reactants and their mixing (consecutively versus all at once.)

2. In this experiment, excess diamine is used to act as an acid acceptor.

3. The student should determine both the viscosity and the melting point of the polymer.

REFERENCES

1. P. W. Morgan and S. L. Kwolek, *J. Chem. Ed.* **36,** 182 (1959).
2. P. W. Morgan, "Condensation Polymers: By Interfacial and Solution Methods," Wiley (Interscience), New York, 1965.
3. P. W. Morgan and S. L. Kwolek, *J. Polym. Sci.* **2,** 181 (1964).
4. E. L. Wittbecker and P. W. Morgan, *J. Polym. Sci.* **40,** 289 (1959).

D

POLYESTERS

The synthetic chemistry for polyesters is briefly reviewed and then is illustrated with a laboratory experiment for preparing poly(1,4-butylene isophthalate) involving the reaction of an acid chloride and a diol.

INTRODUCTION

Polyesters are polymers with repeating carboxylate groups

$$-\overset{\overset{\displaystyle O}{\|}}{C}O-$$

in their backbone chain.

Polyesters are synthesized [1–12] by typical esterification reactions, which can be generalized by the reaction shown in Eq. (1):

$$R\overset{O}{\overset{\|}{C}}-X + N: \rightleftharpoons \left[R-\overset{\overset{\displaystyle \ddot{O}}{|}}{\underset{\underset{\displaystyle X}{|}}{C}}-N \right] \longrightarrow R\overset{O}{\overset{\|}{C}}-N + X: \qquad (1)$$

where N: is a nucleophilic reagent such as $\overline{O}R'$. The rate of reaction is dependent on the structure of R, R′, X, and N and on whether a catalyst is used.

Tartaric acid–glycerol polyesters were reported in 1847 by Berzelius [13] and those of ethylene glycol and succinic acid were reported by Lorenzo in 1863 [14]. Carothers and Van Natta [15] extended much of the earlier work and helped clarify the understanding of the polyesterification reaction in light of the knowledge of polymer chemistry at their time. Polyethylene terephthalate [16, 17] and the polyadipates [18] (for polyurethane resins) were the first major commercial application of polyesters.

The major synthetic methods used to prepare polyesters all involve condensation reactions as shown in Eqs. (2)–(125).

$$\text{HOR—COOH} \xrightarrow{[16]} \text{H}{\left[\text{OR}-\overset{\overset{\text{O}}{\|}}{\text{C}}\right]}_n\text{OH} \quad (2)$$

$$\text{R'COOR—COOH} \xrightarrow{[17]} \text{R'CO}{\left[\text{ORC}\overset{\overset{\text{O}}{\|}}{}\right]}_n\text{OH} \quad (3)$$

$$\text{RCOOROCOR} + \text{R(COOH)}_2 \xrightarrow{[18]} \text{RCO}{\left[\text{ORC}\overset{\overset{\text{O}}{\|}}{}\right]}_n\text{OH} \quad (4)$$

$$\text{HO—R—OH} + \text{R'(COOH)}_2 \xrightarrow{[18,19]} \text{H}{\left[\text{OROCR'C}\overset{\overset{\text{O}\ \ \text{O}}{\|\ \ \|}}{}\right]}_n\text{OH} \quad (5)$$

$$\text{HO—R—OH} + \text{R'(COOR'')}_2 \xrightarrow{[20,20a]} \text{H}{\left[\text{OR}-\overset{\overset{\text{O}}{\|}}{\text{OC}}-\text{R'}\overset{\overset{\text{O}}{\|}}{\text{C}}\right]}_n\text{OH} \quad (6)$$

$$\text{HO—R—OH} + \text{R'(COX)}_2 \xrightarrow{[21,21a]} \text{H}{\left[\text{OROCR'C}\overset{\overset{\text{O}\ \ \text{O}}{\|\ \ \|}}{}\right]}_n\text{X} \quad (7)$$

$$\text{NaOR—ONa} + \text{R'(COX)}_2 \xrightarrow{[22]} \text{Na}{\left[\text{OROCR'C}\overset{\overset{\text{O}\ \ \text{O}}{\|\ \ \|}}{}\right]}_n\text{X} \quad (8)$$
$$(\text{X} = \text{Cl}, \text{OCH}_3)$$

$$\text{HO—R—OH} + \text{R'(CO)}_2\text{O} \xrightarrow{[23]} \text{H}{\left[\text{OROCR'C}\overset{\overset{\text{O}\ \ \text{O}}{\|\ \ \|}}{}\right]}_n\text{OH} \quad (9)$$

$$\text{HO—R—OH} + \text{R'O}\overset{\overset{\text{O}}{\|}}{\text{C}}\text{OR'} \xrightarrow{[24]} \text{H}{\left[\text{OROC}\overset{\overset{\text{O}}{\|}}{}\right]}_n\text{OR'} \quad (10)$$

$$\text{Br—R—Br} + \text{R'(COOAg)}_2 \xrightarrow{[25]} \text{Br}{\left[\text{OROCR'C}\overset{\overset{\text{O}\ \ \text{O}}{\|\ \ \|}}{}\right]}_n\text{Ag} \quad (11)$$

$$\overset{\overset{\text{O}}{\|}}{\underset{\underset{\text{O}}{\rule{2em}{0.4pt}}}{{-}(\text{CH}_2)_n-\text{C}-}} \xrightarrow{[25a]} {\left[\text{O}(\text{CH}_2)_n-\overset{\overset{\text{O}}{\|}}{\text{C}}\right]}_n \quad (12)$$

The use of triols or tricarboxylic acids leads to cross-linked or network polyesters. For example, an alkyd resin is formed by the reaction of glycerol with phthalic anhydride [19].

Some typical examples of the preparation and properties of some representative polyesters are shown in Table 1.

Polyesters [2] find use in fibers [poly(ethylene terephthalate), poly(ethylene oxybenzoate), poly(ester ethers), poly(ester amides), etc.] [1], coatings (especially unsaturated polyesters) [4], plasticizers, adhesives, polyurethane base resins, films, etc. Cross-linked polyesters prepared from glycerol and phthalic anhydride (alkyd resins) have been reviewed [20]. High-melting polyaryl esters have been investigated for high-temperature applications.

Polyesters usually have good thermal and oxidative stability (up to 200°C) but have poor hydrolytic stability at elevated temperatures.

More recent developments involve the preparation of liquid crystalline polyarylates and liquid crystal polyesters in general.

TABLE 1
Preparation of Polyesters by Condensation Reactions

Reactants		Catalyst (g)	Reaction conditions		m.p. (°C)	Mol. wt.	Ref.
Alcohol	Diacid or derivative		Temperature (°C)	Time (hr)			
$HO(CH_2)_nOH$ $n = 2,3,6,10$	Aliphatic type: carbonic, oxalic, succinic, glutaric, adipic, pimelic, and sebacic acids	—	—	—	—	—	a
$HO(CH_2)_nOH$ $n = 2,3,6,10$	Phthalic acid	—	—	—	—	—	a
$HO(CH_2)_4OH$	Isophthaloyl chloride	—	40–218	$\frac{1}{2}$–$\frac{3}{4}$	140–142	—	b
$HO(CH_2)_nOH$ $n = 2,3,6,8,10,18$	Terephthalic acid	—	—	—	—	—	c
$HOCH_2(CF_2)_3CH_2OH$	$(CH_2)_4(COOH)_2$	$ZnCl_2$ (0.01)	150–215	240	Visc. liq.	4000	d
$HOCH_2(CF_2)_3CH_2OH$	$(CF_2)_4(COCl)_2$	—	—	21.5	~35	M_n, 6570	d

[a] W. H. Carothers, U.S. Patent 2,071,053, 1937.
[b] P. J. Flory and F. S. Leutner, U.S. Patent 2,623,934, 1952.
[c] J. R. Whinfield and J. T. Dickson, British Patent 578,079, 1946; E. F. Izard, *J. Polym. Sci.* **8**, 503 (1952); J. R. Caldwell and R. Gilkey, U.S. Patent 2,891,930, 1959.
[d] G. C. Schweiker and P. Robitschek, *J. Polym. Sci.* **24**, 33 (1957).

Jackson and co-workers reported in 1976 [21,22] (on work carried out in 1971) that poly-*p*-oxybenzoyl-modified poly(ethylene terephthalate) was found to exhibit thermotropic liquid crystalline characteristics. A variety of thermotropic liquid crystalline polyarylates have since been described, and some of these also contain flexible "spacer" units [23–30].

$$CH_3O\overset{O}{\overset{\|}{C}}{-}\underset{}{\bigcirc}{-}\overset{O}{\overset{\|}{C}}OH + HOCH_2CH_2OH + HO\overset{}{\overset{\|}{C}}{-}\underset{}{\bigcirc}{-}\overset{}{\overset{\|}{C}}OH \longrightarrow$$

$$\left[\left(-O-\bigcirc-\overset{}{\underset{O}{\overset{\|}{C}}}-\right)_n\!\!\left(O-CH_2CH_2O\overset{}{\underset{O}{\overset{\|}{C}}}-\bigcirc-\overset{}{\underset{O}{\overset{\|}{C}}}-\right)_n\right]_x \quad (13)$$

The field of liquid crystal polyesters (LCP) is expanding very rapidly and has been reviewed [22]. The area of chiral liquid-crystalline polyesters has also been researched [32]. Several companies have started to exploit their use on a commercial scale. The high-melting LCP's are expected to find use as replacements for metals, ceramics, composites, etc., in such areas as telecommunications equipment, aerospace, automotive parts, computers, and electrical/electronic components.

The total world market for liquid crystal polymers has been estimated [32] to be 6 million pounds with a possibility for growth in excess of 13% annually over the next 10-year period.

Polyesters are essentially prepared by the typical procedures used for the preparation of the monoesters described in an earlier volume by the present authors [33]. Catalysts are used to increase the rate of esterification. Aromatic acids and aromatic alcohols afford high-melting polyesters as compared to viscous liquids prepared from aliphatic starting materials. The use of stoichiometric amounts affords high molecular weight products whereas an excess of one reactant lowers the molecular weight. In many cases, hydroxy acids have a great tendency toward forming cyclic dimers, especially if five- or six-membered rings can be formed, as in the case of hydroxyacetic acid [34].

$$2\text{HOCH}_2\text{COOH} \longrightarrow \text{(glycolide)} + 2\text{H}_2\text{O} \qquad (14)$$

Heating glycolide with zinc chloride affords a linear polyester; $\text{HO(CH}_2\text{COO})_n\text{H}$.

The polycondensation of diols with diacid chlorides is effected either by directly heating the components in a nitrogen atmosphere or by a variation of the Schotten–Baumann reaction using trialkylamine dissolved in an inert solvent (acetone) [35]. The latter reaction can also be carried out interfacially by using a bisphenol dissolved in aqueous base (NaOH) and then adding the diacid dichloride in a water-immiscible solvent. Using the latter method, yields are approximately 80–100% and the molecular weights are as high as 80,000–90,000. The interfacial polycondensation procedure gives higher molecular weight polyesters than the melt polymerization procedure. Monofunctional phenols or acid chlorides act as molecular weight regulators.

REFERENCES

1. J. M. Hawthorne and C. J. Heffelfinger, *Encycl. Polym. Sci. Technol.* **11**, 1 (1969).
2. I. Goodman, *Encycl. Polym. Sci. Eng.* **12**, 1 (1988).
3. I. Goodman, *Encycl. Polym. Sci. Technol.* **11**, 62 (1969).
4. H. V. Boening, *Encycl. Polym. Sci. Technol.* **11**, 29 (1969).
5. V. V. Korshak and S. V. Vinogradova, "Polyesters." Elsevier, Amsterdam, 1953.
6. V. V. Korshak and S. V. Vinogradova, "Polyesters." Pergamon, Oxford, 1969.
7. R. Hill and E. E. Walker, *J. Polym. Sci.* **3**, 609 (1948).
8. H. J. Hagemeyer, U.S. Patent 3,043,808, 1962.
9. H. Batzer and F. Wiloth, *Makromol. Chem.* **8**, 41 (1952).
10. B. M. Grievson, *Polymer* **1**, 499 (1960).
11. M. J. Hurwitz and E. W. Miller, French Patent 1,457,711, 1966.
12. Borg-Warner Corp., British Patent 1,034,194, 1966.
13. J. Berzelius, *Rapp. Annu.* **26**, 1 (1847); *Jahresbericht* **12**, 63 (1833).
14. A. V. Lorenzo, *Ann. Chim. Phys.* **67**(2), 293 (1863).
15. W. H. Carothers and E. J. Van Natta, *J. Am. Chem. Soc.* **55**, 4714 (1933).
16. J. R. Whinfield, *Nature (London)* **158**, 930 (1946).
17. S. K. Agarawal, *Ind. Chem. J.* **5**, 25 (1970).
18. O. Bayer, *Justus Liebigs Ann. Chem.* **549**, 286 (1941).
19. M. Callahan, U.S. Patents 1,191,732 and 1,108,329-39, 1914.
20. R. G. Mraz and R. P. Silver, *Encycl. Polym. Sci. Technol.* **1**, 663 (1964).
21. W. J. Jackson, Jr., and H. F. Kuhfuss, *J. Polym. Sci. Polym. Chem. Ed.* **14**, 2043 (1976).
22. W. J. Jackson, Jr., *Mol. Cryst. Liq. Cryst.* **169**, 23 (1989).
23. A. Blumstein, ed., "Liquid Crystalline Order in Polymers." Academic Press, New York, 1978.
24. A. Ciferri, W. R. Krigbaum, and R. Meyers, eds., "Polymer Liquid Crystals." Academic Press, New York, 1982.
25. L. L. Chapoy, ed., "Recent Advances in Liquid Crystalline Polymers. Elsevier, New York, 1985.
26. S. Antoun, R. W. Lenz, and J.-I. Jin, *J. Polym. Sci. Polym. Chem. Ed.* **19**, 1901 (1981).

27. A. Blumstein, ed., "Polymeric Liquid Crystals," Plenum, New York, 1985.
28. C. K. Ober, J.-I. Jin, and R. W. Lenz, *Adv. Polymer. Sci.* **59,** 103 (1984).
29. J. L. White, *J. Appl. Polym. Sci., Appl. Polym. Symp.* **41,** 3 (1985).
30. N. Koidi, *Mol. Cryst. Liq. Cryst.* **139,** 47, (1986).
31. E. Chiellini and G. Galli, *Faraday Discussions Chem. Soc.* **79,** 241 (1985).
32. Chem. Systems' Eighteenth Annual PERP Seminar Program, jointly with Chem. Systems' Third Annual PPE Seminar Program, January 17–18, 1990. Four Seasons Hotel, Houston Center, Houston, TX.
33. S. R. Sandler and W. Karo, "Organic Functional Group Preparations," Vol. 1, Second Ed. Chapter 10. Academic Press, New York, 1983.
34. C. A. Bischoff and P. Walden, *Chem. Ber.* **26,** 262 (1893); *Justus Liebigs Ann. Chem.* **279,** 45 (1894).
35. G. S. Papava, N. A. Maisuradze, S. V. Vinogradova, V. V. Korshak, and P. D. Tsiskarishvili, *Soobshch. Akad. Nauk. Gruz. SSR* **62,** 581 (1971); *Chem. Abstr.* **75,** 77340s (1971).

EXPERIMENT 9

Preparation of Poly(1,4-butylene isophthalate)

INTRODUCTION

Details of this experiment may be found in Ref. 1.

SAFETY PRECAUTIONS

Before this experiment is carried out, the student must read the material safety data sheets (MSDS) for all the chemicals used as well as for the products. The instructor must approve that you have read and understood the MSDS for the safe handling of these materials.

Please be advised that all chemicals should be considered hazardous and should be handled in a hood and with proper personal protective equipment (lab coat, proper gloves, approved safety glasses, and/or goggles). Avoid inhaling vapors and/or aerosolized materials. Avoid skin/eye contact with all chemicals at all times. Wash hands frequently. See the instructor if you have any questions or concerns.

APPARATUS

1. A 50-ml glass ampoule used for polymerization equipped with a septum with a syringe needle for the inlet of nitrogen and extending below the surface of the reactants and another syringe needle to go above the reactants to act as an exit for the HCl that is generated in this reaction
2. A heating bath or electric heating mantle capable of heating the tube to 218°C
3. Balance

REAGENTS AND MATERIALS

1. Isophthaloyl chloride (isophthaloyl dichloride)
2. 1,4-Butanediol
3. Nitrogen

PROCEDURE

$$HOCH_2CH_2CH_2CH_2OH + ClC(O)\text{-}C_6H_4\text{-}C(O)Cl \xrightarrow{-2HCl}$$

$$\left[-O(CH_2)_4OC(O)\text{-}C_6H_4\text{-}C(O)- \right]_n$$

1. Add 6.3 g (0.0310 mol) of isophthaloyl chloride and 2.8 g (0.0311 mol) of 1,4-butanediol (1% excess) to a 50-ml ampule used for polymerization equipped with a capillary tube for a nitrogen inlet extending below the surface of the reaction mixture.
2. Add the reactants in a nitrogen atmosphere, and slowly continue the nitrogen flow during the reaction. The reaction is exothermic and warms up to 40°C.
3. After 10 min, heat the reaction mixture to 218°C and keep there for 35 min to afford a clear viscous polymer with a melt viscosity of 2500 poises. A clear film of the amorphous material hardens on standing to form a white opaque polymer, m.p. 140–142°C (see note).

NOTE

The student should measure the melt viscosity and the softening melting point of the polymer.

REFERENCE

1. P. J. Flory and F. S. Leuther, U.S. Patents 2,623,034 and 2,589,688, 1952.

EPOXY RESINS

The synthesis chemistry of epoxy resins is briefly reviewed, along with a listing of numerous references from the literature. The curing of an epoxy resin by reaction with a polyamine is described in the following experiment. The reaction can be followed by running either NMR or infrared spectra at periodic intervals. After standing for several hours or days, it is analyzed again to see if there was any further change. A comparison of two different but related amines is made to see how amine functionality affects the rate of reaction, all run at room temperature.

INTRODUCTION

Epoxy resins are usually prepared from compounds (or polymers) containing two or more epoxy groups that have been reacted with amines, anhydrides, or other groups capable of opening the epoxy ring and forming thermosetting products. Polymers from monoepoxy compounds have already been described in Sandler and Karo [1].

Schlack [2] and Castan [3,4] are credited with the earliest U.S. patents describing epoxy resin technology. Greenlee [5] further emphasized the use of bisphenols and their reaction with epichlorohydrin to yield diepoxides capable of reaction with crude tall oil resin acids to yield resins useful for coatings. The use of diepoxide resins that are cured with amines was reported by Whittier and Lawn [6] in a U.S. patent in 1956.

The introduction of epoxidation techniques for polyunsaturated natural oils by Swern and colleagues [7,8] led to industrial interest in the preparation of epoxy compounds useful for resin production [9,10].

Epoxy resin technology has been reviewed and a number of relevant references are available [7,11–16].

The applications of epoxy resins are not only used for adhesives but in the area of coatings used for appliances, automobiles, and cans. These resins have the added advantage of being solventless systems, which helps avoid air pollution problems in plants. The pattern of consumption of epoxy resins is described in Table 1.

The growth in production of epoxies is up from 183 million pounds in 1972 [17], to an estimated sales of epoxy products of 464 million pounds in 1990 [18]. Epoxy thermosetting resin sales in 1990 are also reported to be 499 million pounds [19].

The approximate epoxy resin production capacity as of 1993 was more than 800 million pounds per year. Major producers and approximate capacities are tabulated in Table 2. Ciba-Geigy, Dow, and Shell are the major producers.

SAFETY PRECAUTIONS

Epoxy resins and their curing agents are considered primary skin irritants. Contact with epoxy resins should only be made using gloves and face shields, and while working in hoods or well-ventilated areas [20]. Some individuals, on prolonged contact with epoxy resins, may develop a skin sensitization evidenced by blisters or other dermatitis conditions. Other individuals may develop an asthma-like condition. Contaminated gloves and work clothes should be changed immediately and either laundered or discarded. Contaminated shoes should be discarded. Frequent washing of hands is advisable and strict personal hygiene must be practiced. Some aromatic amine curing agents may be carcinogenic. All amine curing agents or N-heterocyclic types should be suspect and handled with great care. A properly cured epoxy resin system usually presents no health problems

TABLE 1
Epoxy: Pattern of Consumption[a]

	1000 metric tons		
Market	1972	1973	1990[c]
Bonding and adhesives[b]	5.1	5.4	13
Flooring, paving, and aggregates	4.6	5.4	12
Protective coatings			
Appliance finishes	2.9	3.8	
Auto primers	5.6	6.7	
Can and drum coatings	9.5	10.9	89
Pipe coatings	2.3	2.2	
Plant maintenance	7.4	8.7	
Other (including trade sales)	9.6	12.1	
Reinforced plastics[d]			
Electrical laminates	6.2	9.1	25
Filament winding	3.4	3.8	—
Other	2.9	3.0	14
Tooling, casting, and molding	7.8	7.9	13
Exports	9.1	12.2	31
Other	6.6	7.9	15
Total	83.0	99.1	212

[a] Data for 1972 and 1973 reprinted from *Mod. Plast.* **51,** 39 (1974). Copyright 1974 by Modern Plastics. Reprinted with permission of the copyright owner.
[b] Includes flooring, road coating, and TV resins.
[c] Data estimated for 1990 found in *Mod. Plast.* **21,** 56 (1981).
[d] Does not include reinforcements.

TABLE 2
Epoxy Resin Production

Producer	Approximate production capacity for 1993 in million pounds per year[a]
Shell Chemical Co.	290
Dow Chemical Co.	265
Ciba-Geigy Co.	190
Union Carbide Corp.	15
Rhone Poulenc	30
Reichhold Chemical Co.	12

[a] From W. F. Stahl, "Chemical Economics Handbook," Marketing Research Report on Epoxy Resins, p. 580.0600A, August, 1994.

relating to skin irritation. Individuals who show sensitivity should discontinue handling epoxy compounds, as hypersensitivity will develop making it difficult even to come close to those materials without developing dermatitis or other reactions. For additional safety information, see Refs. 16 and 21.

ANALYSIS OF EPOXY RESINS

Epoxy resins are analyzed for epoxy or oxirane content, which is reported as the epoxy or oxirane equivalent or epoxy equivalent weight, i.e., the weight of resin in grams that contains a 1-g equivalent of an epoxy group. The "epoxy value" designates the fractional number of epoxy groups per 100 g of epoxy resin. Percentage oxirane oxygen is used for epoxidized oils and dienes.

Analytically, epoxy groups are determined by the reaction with hydrogen halide and back titration with a standard base. Other functional groups present may cause interference problems and result in poor end points. Pyridinium chloride–pyridine is a recommended reagent for the analysis of bisphenol–diglycidyl ether resins [22,23].

Other improved analytical procedures allow direct titration of the epoxy group [23]. This is achieved by the use of hydrogen bromide dissolved in an anhydrous protic solvent such as glacial acetic acid [24]. The methods developed by Jay [25] and Dijkstra and Dahmen [26] using quaternary ammonium bromide or iodide in acetic and perchloric acid solutions for titration are considered the best general techniques for a wide variety of epoxides (hindered and unhindered alicyclic epoxides). The use of the halogen acid procedure fails for epoxides that undergo intramolecular rearrangement to aldehydes or ketones [27].

More recently, Eggers and Humphrey [28] have reported on the application of gel permeation chromatography to monitor epoxy resin molecular distributions and curing. The preparation given describes an infrared spectrophotometric method to follow curing.

CONDENSATION–ELIMINATION REACTIONS

Epoxy Compounds via Epichlorohydrin

Epoxy resins are generally prepared by the reaction of epichlorohydrin with active hydrogen-bearing compounds:

$$\text{Cl}-\text{CH}_2\text{CH}-\text{CH}_2 + \text{R}-\text{XH} \xrightarrow[-\text{HCl}]{\text{base}} \text{RX}-\text{CH}_2-\text{CH}-\text{CH}_2 \qquad (1)$$
$$\underset{\text{O}}{\diagdown\diagup}$$

X = O, S
R = aliphatic or aromatic

The reaction involves a chlorohydrin intermediate, which is then treated with a base to give the resulting epoxy compound.

CURING–POLYMERIZATION REACTIONS OF EPOXY COMPOUNDS AND RESINS

Epoxy resins are cured [29] by reaction of the epoxy group with other functional groups to give linear, branched, or cross-linked products as described in Eq. (2) and in Table 3.

$$\text{Z} + \text{CH}_2-\text{CHR} \longrightarrow \text{Z}^+-\text{CH}_2-\text{CH}-\text{R} \qquad (2)$$

$\text{Z} = \text{R}_3\text{N}, \text{ROH}, \text{RCOOH}, (\text{RCO})_2\text{O}, \text{RNH}_2, \text{RCONH}_2, \text{RSH}$, etc.

Compounds Z are active hydrogen compounds such as amines, anhydrides, and acids [30]. The curing reaction can also involve homopolymerization catalyzed by Lewis acids or tertiary amines. In most cases, reactions catalyzed by

TABLE 3
Reactivity of Amine Accelerators with Anhydride–Epoxy Resin Systems[a]

Accelerator	Gel time at 65°C, 100-g samples[b]		
	A (hr)	B (hr)	C (hr)
Pyridine	1	0.5	1.5
Diethylamine	—	9	42
Triethylamine	5	2.5	4.5
α-Methylbenzyldimethylamine	5	2.5	4.5
Dibutylaminopropylamine	5	2.5	1.5
Diethylaminopropylamine	5	3.5	3.5
Dibutylaminopropylamine acetate	4	4	6.5
p,p'-Methylenebis(N,N-dimethylaniline)	1	1	1
2-Aminopyridine	3	1.5	2.5
2-Amino-3-methylpyridine	3	2	3.5
2-Amino-4-methylpyridine	16	—	7
2-Amino-5-methylpyridine	17	2.5	3.5
2-Amino-6-methylpyridine	4	5.5	8.5
2,6-Diaminopyridine	8	21	—

[a] Reprinted from G. M. Kline, *Mod. Plast.* **42**, 152 (1964). Copyright April, 1964 by Modern Plastics. Reprinted by permission of the copyright owner.
[b] Formulations: Araldite 6010 plus 2% of the indicated accelerator and (A) 102.4 phr hexahydrophthalic anhydride : HET anhydride 50 : 50, (B) 93.9 phr hexahydrophthalic anhydride : HET anhydride 60 : 40, and (C) 85.5 phr hexahydrophthalic anhydride : HET anhydride 70 : 30.

E. Epoxy Resins

Lewis acids are too fast for practical systems, whereas reactions catalyzed by tertiary amines are too slow and too highly temperature and concentration dependent to make them reliable as the sole curing agent.

It is interesting to note that the activity of tertiary amines toward epoxide does not always correlate with their basicity but also depends strongly on steric factors. For example, triethylamine appears more reactive than tributylamine or benzyldiethylamine [31]. However, all epoxy groups are of equal reactivity. Electron-attracting groups increase the rate with nucleophilic agents (amines, inorganic bases), whereas electron-donating groups (methylene, vinyl) improve reactivity with acidic-type electrophilic) curing agents. For example, internal epoxy groups favor reaction with electrophilic curing agents [32].

Some of the typical curing reactions are illustrated in Scheme 1 [33–43].

The choice of the specific curing agent is based on end-use considerations with regard to desired properties of the cured resin. The amount of curing agent, if of the active hydrogen type, is calculated as one epoxy group for each active hydrogen of the reagent. Too much curing agent may cause embrittlement. Tertiary amines and Lewis acids are required only in catalytic amounts such as 5–15 phr.

With the diglycidyl derivative of bisphenol A, aromatic amines such as 4,4'-methylene dianiline or diaminodiphenyl sulfone provide good thermal stability for the final cured resin. Although aliphatic primary amines react more rapidly (triethylenetetramine cures the above epoxy resin based on bisphenol A in 30 min at room temperature and causes it to exotherm up to 200°C), they are more difficult to handle and offer poor thermal stability.

Anhydrides generally provide pot lives of days or months and cure usually at 100–180°C with very little exotherm. Tertiary amines accelerate the time for gelation but still require the elevated temperature cure to obtain optimum properties. Anhydrides usually give brittle products but the addition of polyether flexibilizing groups yields more elastomeric products [44].

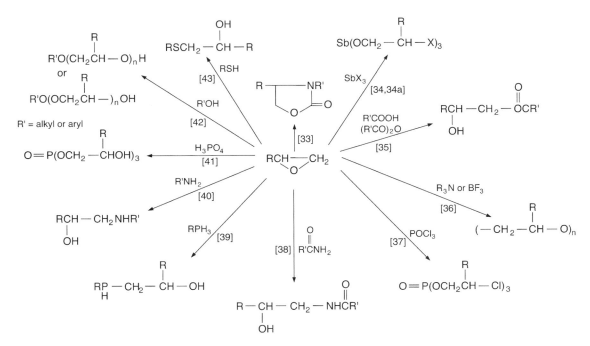

Scheme 1 Epoxy resin-curing reactions.

REFERENCES

1. S. R. Sandler and W. Karo, *Polymer Syntheses,* Vol. I, 2nd ed., Boston: Academic Press (1992).
2. P. Schlack, U.S. Patent 2,131,120 (1938); German Patent 676,117 (1939).
3. P. Castan, U.S. Patent 2,324,483 (1943).
4. P. Castan, Swiss Patent 211,114 (1940).
5. S. O. Greenlee, U.S. Patent 2,493,486 (1950).
6. F. Whittier and R. J. Lawn, U.S. Patent 2,765,288 (1956).
7. D. Swern, *Chem. Rev.* **45,** 1 (1949).
8. T. W. Findley, D. Swern, and J. T. Scanlan, *J. Am. Chem. Soc.* **67,** 412 (1945).
9. B. Phillips, "Peracetic Acid and Derivatives." Bulletin P-58-0283. Union Carbide Chem. Co., New York, New York, 1958; Bulletin 4. Food Machinery and Chemical Corp., Becco Chem. Div., 1952 (revised 1957); R. J. Gall and F. P. Greenspan, *Ind. Eng. Chem.* **47,** 147 (1955); Bulletin P61-454. E. I. du Pont de Nemours & Co., Inc., 1954; Bulletin A6282. E. I. du Pont de Nemours & Co., Inc., 1955.
10. F. P. Greenspan and R. J. Gall, *J. Am. Oil Chem. Assoc.* **33,** 391 (1956).
11. S. R. Sandler and W. Karo, *Organic Functional Group Preparations,* Vol. 1, pp. 99–115, Academic Press, New York, 1968.
12. H. Lee and K. Neville, "Epoxy Resins, Their Applications and Technology," McGraw-Hill, New York, 1957; *Handbook of Epoxy Resins.* New York: McGraw-Hill, 1967; *Encycl. Polym. Sci. Technol.* **6,** 209 (1967); I. Skeist, *Epoxy Resins.* Princeton, New Jersey: Van Nostrand-Rheinhold, 1958; M. W. Ranney, *Epoxy and Urethane Adhesives.* New Jersey: Noyes Data Corp., 1971; G. R. Somerville and P. D. Jones, *Abstr. 168th Am. Chem. Soc. Meet., Atlantic City, N.J., 1974* Abstract ORPL, 146 in the Organic Coatings and Polymer Chemistry Division (1974).
13. L. V. McAdams and J. A. Gannon, *Encyclopedia of Polymer Science & Engineering* (2nd ed.) **6,** 322 (1986).
14. J. Gannon, Kirk-Othmer Encyclopedia of Chemical Technology, (4th ed), **9,** 730 (1994).
15. W. F. Stahl, *Chemical Economics Handbook (CEH)*, Marketing Research Report on Epoxy Resins, p. 580.0600A, August (1994).
16. F. Lohse, *Makromol. Chem., Macromol. Symp.* **7** (1987), *Chem. Abstr.* **106,** 103051d (1987).
17. W. F. Fallwell, *Chem. Eng. News* **51**(40), 8 (1973).
18. *Modern Plastics International* **21,** 55 (1991).
19. *Chemical & Engineering News* **70**(26), 38 (1992).
20. "Industrial Hygiene Bulletin on Handling Epon Resins and Auxiliary Chemicals," SC 62-33. Shell Chem. Corp., 1962.
21. J. E. Berger, K. I. Darmer, and E. F. Phillips, *High Solids Coat.* **16** (Dec. 1980).
22. D. W. Knoll, D. H. Nelson, and P. W. Keheres, *Pap., 134th Am. Chem. Soc. Meet., Chicago, 1958* Division of Paint, Plastics, and Printing Ink Chemistry, Paper No. 5 p. 20 (1958).
23. B. Dobinson, W. Hoffmann, and B. P. Stark, *The Determination of Epoxide Groups.* Oxford: Pergamon, 1969.
24. A. J. Durbetaki, *Anal. Chem.* **28,** 2000 (1956).
25. R. R. Jay, *Anal. Chem.* **36,** 667 (1964).
26. R. Dijkstra and E. A. M. F. Dahmen, *Anal. Chim. Acta* **31,** 38 (1964).
27. A. J. Durbetaki, *Anal. Chem.* **29,** 1666 (1957).
28. E. A. Eggers and J. S. Humphrey, Jr., *J. Chromatogr.* **55,** 33 (1971).
29. H. Lee and K. Neville, "Handbook of Epoxy Resins," Chapter 5. McGraw-Hill, New York, 1967.
30. "Epon Resins," Technical Brochure SC 71-1. Shell Chem. Corp., Houston, Texas, 1971.
31. L. Schechter and J. Synstra, *Ind. Eng. Chem.* **48,** 94 (1956).
32. R. F. Fischer, *Ind. Eng. Chem.* **52,** 321 (1960) (refer to Scheme I).
33. Hercules Powder Co., British Patent 892,361 (1962).

34. F. P. Greenspan, *High Polym.* **19,** 152–172 (1964); R. W. Rees, U.S. Patent 3,050,507 (1962); J. W. Pearce and J. Kana, *J. Am. Oil Chem. Soc.* **34,** 57 (1957); R. J. Gall and F. P. Greenspan, *ibid.* p. 161; W. Wood and J. Termini, *ibid,* **35,** 331 (1958); F. P. Greenspan and R. E. Light, Jr., Canadian Patent 560,690 (1958).
34a. A. F. Chadwick, D. O. Barlow, A. A. D'Addieco, and J. G. Wallace, *J. Am. Oil Chem. Soc.* **35,** 355 (1958).
35. W. D. Niederhauser and J. E. Koroly, U.S. Patent 2,485,160 (1949); H. Fukutani, M. Tokizawa, H. Okada, and N. Wakabayashi, German Offen. 2,143,071 (1972); *Chem. Abstr.* **77,** P20675S (1972).
36. W. F. Pfohl, U.S. Patent 5,003,066 (1991).
37. T. W. Evans and E. C. Shokal, U.S. Patent 2,599,817 (1952).
38. K. Okamoto, Japanese Patent JP 62124111A2 (1987); *Chem. Abstr.* **108,** 151530x (1987).
39. Data taken from "Preparation of O-Type Polymer (III) Solution Copolymerization of Glycidyl Methacrylate," Tech. Bull. 6J12 1000A. Nippon Oil & Fats Co., Ltd., Tokyo, Japan 1974.
40. E. M. Gotlib, O. M. Voskresenskaya, L. V. Verizhnikov, A. G. Liakumovich, and P. A. Kirpichnikov, *Vysokomol. Swedin Ser. A* ***33(6),*** 1192 (1991); *Chem. Abstr.* **115,** 115577d (1991).
41. L. S. Corley, U.S. Patent 4,581,436 (1986); *Chem. Abstr.* **105,** 61570 (1986).
42. L. S. Corley, U.S. Patent 5,579,931 (1986); *Chem. Abstr.* **105,** 79939f (1986).
43. K. Okamoto, Japanses Patent JP62132914A2 (1987). *Chem. Abstr.* **107,** 237990p (1987).
44. S. R. Sandler and F. R. Berg, U.S. Patent 3,437,671 (1969).

EXPERIMENT 10

Preparation of a Cured Epoxy Resin by the Room Temperature Reaction of Bisphenol A Diglycidyl Ether with Polyamines

SAFETY PRECAUTIONS

Before this experiment is carried out, the student must read the material safety data sheets (MSDS) for all the chemicals used as well as for the products. The instructor must approve that you have read and understood the MSDS for the safe handling of these materials.

Please be advised that all chemicals should be considered hazardous and should be handled in a hood and with proper personal protective equipment (lab coat, proper gloves, approved safety glasses, and/or goggles). Avoid inhaling vapors and/or aerosolized materials. Avoid skin/eye contact with all chemicals at all times. Wash hands frequently. See the instructor if you have any questions or concerns.

APPARATUS

1. Aluminum weighing dishes
2. Test tubes
3. Salt plates for infrared (IR) spectrophotometer
4. Nuclear magnetic resonance (NMR) tubes
5. Deuterated dimethyl sulfoxide for NMR
6. IR spectrophotometer
7. NMR
8. Balance
9. Spatula

REAGENTS AND MATERIALS

1. Poly(bisphenol A - co- epichlorohydrin)-glycidyl end capped- CAS [25036-25-3], av. mol. wt. 348, m.p. 41–44°C; d 1.169.

2. Diethylenetriamine, 99%, CAS[111-40-0], $H_2NCH_2CH_2NHCH_2CH_2NH_2$, mol. wt. 103.17, m.p. $-35°C$, d 0.955.
3. Triethylenertetramine, 60%, CAS[112-24-3], $H_2NCH_2CH_2NHCH_2CH_2NHCH_2CH_2NH_2$, mol. wt. 146.24, m.p. 12°C, d 0.982.

PROCEDURE

1. To a test tube or aluminum weighing dish add 10.44 g (0.03 mol or 0.06 equiv.) of the Bisphenol A epoxy resin described above and then 2.06 g (0.02 mol or 0.06 equiv.) of diethylenetriamine.
2. Mix well with a spatula.
3. Quickly place a small amount of material between two IR salt plates and run the IR immediately (time = 0). Run the IR spectrum every 15 min by scanning only the absorption peaks at 3226 and 912 cm^{-1} to follow the reaction.
4. Make a plot of data to follow the rate of reaction with time.
5. Store samples until the next laboratory period and run the IR again to see if there is any difference. Record the results.
6. Run the same experiment side by side using triethylenetetramine (use 6.49 g epoxy resin or 0.02 mol/0.04 equiv. and 1.46 g of the amine or 0.01 mol/0.04 equiv.) and record the IR in a similar fashion.

REPORT

1. From the infrared plots of experimental data, determine which amine is more reactive. Also describe if there was an exotherm and the maximum temperature reached during the course of your observation. Describe the reactions taking place and use the IR to back up your suggestions.
2. The NMR can also be used if desired.
3. What will you be looking for in your analysis?

NOTES

1. The infrared spectra of the epoxy resin, diethylenetriamine, and triethylenetetramine should be run alone as controls.
2. To measure the rate of the reaction, see Experiment 4 in Section II.

POLYMERIZATION OF VINYL ACETATE

INTRODUCTION

The section on acrylic esters discusses in great detail many aspects and techniques that are of importance for free radical polymerizations. Most of these can be applied to other monomer systems with minor modifications that are specific to the particular monomer whose polymerization is under consideration.

For example, an aqueous suspension homopolymerization would normally not be appropriate for water-soluble monomers such as acrylamide or acrylic acid. For such cases, it may be possible that with high levels of neutral salts, the water solubility of such monomers may be reduced to permit the use of suspension polymerization techniques. Other monomers may have other unique properties that require some modifications of the basic procedures.

Polymers of vinyl acetate are produced commercially in large volumes. Emulsions of poly(vinyl acetate) are used as paint bases, coatings, and adhesives. Overall, the polymerization processes involving vinyl acetate are quite straightforward.

However, consider some of the physical properties shown in Table 1.

From Table 1, note that the boiling point is close to the temperature at which many vinyl polymerizations are often run (80°C). In fact, the water–vinyl acetate azeotrope boils even lower. Therefore, emulsion polymerizations have to be initiated at a moderate temperature.

With the density of this monomer being less than that of water, stirring is required during suspension or emulsion polymerizations to prevent the formation of a layer of monomer on top of the aqueous system.

Note also that the monomer is moderately soluble in water. It has a tendency to hydrolyze to form a small amount of acetic acid. It is therefore desirable to have salts present in the aqueous phase and to buffer the solution.

TABLE 1
Selected Physical Properties of Vinyl Acetate[a]

Boiling point	72.7°C/760 mm Hg
Density, d_4^{20} (g/ml)	0.9312
Solubility in water	
At 29°C	2.2 wt%
At 80°C	3.0 wt%
Boiling point of azeotrope with water	66°C
Heat of polymerization (kcal/mol)	21.0 ± 0.5

[a] Taken from S. R. Sandler and W. Karo, "Polymer Syntheses, 2nd Ed., Vol. III, p. 206, Academic Press, San Diego, 1996.

As the heat of polymerization is quite high, sufficient water has to be present to control the temperature that may evolve during the process.

In Experiment 12, a small amount of a buffering salt is used. By use of a mixture of monomers, some of which have a high boiling point, a monomer system is used that can be maintained at a reasonably high reaction temperature. By use of a gradual monomer addition technique, the exothermic reaction is controlled.

EXPERIMENT 11

Seeded Emulsion Terpolymerization of Vinyl Acetate, Butyl Acrylate, and Vinyl Neodecanoate with Gradual Monomer and Initiator Additions

PRINCIPLE

Emulsion polymerizations lend themselves to many procedural variations. In so-called "seeded" procedures, a small quantity of the monomer is polymerized by an emulsion technique. According to the theory of emulsion polymerization, the surfactant concentration in the substrate fixes the number of particles formed. When more monomer (within broad limits) is added to such an emulsion, the existing particles pick up the new monomer and become larger in diameter. The relative surface areas of small particles are larger than those of larger particles. As the rate of absorption is directly related to the surface area, if the seed latex consists of particles of a range of diameters, the addition of the fresh monomer will tend to produce particles of a more uniform size distribution. On the basis of geometry, to double the diameter of a particle, seven times the volume of the initial particle is required, if we ignore the shrinkage factor when a monomer is polymerized. This means that a sizable amount of fresh monomer may be added to swell seed particles before there is a very noticeable change in the overall diameters of the product.

The polymerization of a mixture of more than one monomer leads to copolymers if two monomers are involved and to terpolymers in the case of three monomers. At low conversions, the composition of the polymer that forms from just two monomers depends on the reactivity of the free radical formed from one monomer toward the other monomer or the free radical chain of the second monomer as well as toward its own monomer and its free radical chain. As the process continues, the monomer composition changes continually and the nature of the monomer distribution in the polymer chains changes. It is beyond the scope of this laboratory manual to discuss the complexity of reactivity ratios in copolymerization. It should be pointed out that the formation of terpolymers is even more complex from the theoretical standpoint. This does not mean that such terpolymers cannot be prepared and applied to practical situations. In fact, Experiment 5 is an example of the preparation of a terpolymer latex that has been suggested for use as an exterior protective coating.

COMPOSITION

In the case of copolymerizations, the composition of the initially formed copolymer is often quite different than that of the charged monomer composition.

The composition of the initial copolymer from a solution of two monomers that are capable of copolymerization is given by

$$M_0 = \frac{(P_0 - 1) + \sqrt{(1 - P_0)^2 + 4P_0 r_1 r_2}}{2r_1}$$

where M_0 is the mole ratio of monomer$_1$ to the charged monomer$_2$, P_0 is the mole ratio of the *initially* formed copolymer; $r_1 r_2$ is the product of the reactivity ratios of the two monomers, and r_1 is the reactivity ratio of monomer$_1$.

If the final desired composition of a copolymer, P_t, is set equal to P_0, an M_0 can be calculated.

If this composition, M_0, is allowed to polymerize to a low conversion as a seed particle, a polymer of composition, P_t, is formed. If an additional monomer is now added gradually at the same rate as the polymer is formed, a monomer composition, M_0, is maintained and only a polymer of composition, P_t, is formed.

The indicated procedure for the production of a copolymer of predetermined final composition is to initiate a small amount of seed composition, M_0, followed by the gradual addition of the monomers of the final composition, P_t. If the total charge of the addition monomer, P_t, greatly exceeds the composition of M_0 and the polymerization is stopped before all of the available monomer has been used up, the final product will consist substantially of composition P_t.

Another aspect of seeded polymerization is that a seed may be formed from one monomer composition whereas the added monomer may be of a different composition. This may lead to "core-shell" latex particles. Such copolymers may have substantially different properties then "regular" copolymers.

SAFETY PRECAUTIONS

Before this experiment is carried out, the student must read the material safety data sheets (MSDS) for all the chemicals used as well as for the products. The instructor must approve that you have read and understood the MSDS for the safe handling of these materials.

Please be advised that all chemicals should be considered hazardous and should be handled in a hood and with proper personal protective equipment (lab coat, proper gloves, approved safety glasses, and/or goggles). Avoid inhaling vapors and/or aerosolized materials. Avoid skin/eye contact with all chemicals at all times. Wash hands frequently. See the instructor if you have any questions or concerns.

APPARATUS

1. Laboratory fume hood
2. 1-liter four-necked round-bottom flask
3. Two Claisen-type adapters
4. Reflux condenser
5. 250-ml addition funnel
6. 50-ml addition funnel

7. Mechanical stirrer with stainless-steel stirring rod and blade and speed adjustment to permit a stirring rate in the range of 150–200 rpm
8. Glascol heater and powerstat
9. Thermometer
10. Nitrogen inlet tube
11. Funnels
12. A 200 mesh stainless-steel screen with provisions for use as a filter
13. Vinyl gloves, apron, face shield, and goggles
14. Miscellaneous laboratory equipment

REAGENTS AND MATERIALS

1. Deionized water
2. 120 g of vinyl acetate
3. 30.6 g of butyl acrylate
4. 49.4 g of vinyl neodecanoate (VYNATE Neo-10 monomer from Union Carbide, Danbury, CT)
5. 2.0 g Cellosize hydroxethyl cellulose WP-300
6. 2.0 g Tergitol NP-40
7. 2.6 g Tergitol NP-15
8. 2.20 g Siponate DS-4
9. 0.40 g ammonium bicarbonate
10. 0.56 g ammonium persulfate
11. Nitrogen

PROCEDURE [1]

1. In a hood, add 136 ml of deionized water to a 1-liter four-necked flask equipped with two Claisen-type adapters that accommodates a reflux condenser, a mechanical stirrer, a thermometer, a 250-ml addition funnel, a nitrogen inlet tube, and a 50-ml addition funnel.
2. Then, with a moderate stirring speed, add 4.00 g of Cellosize hydroxyethyl cellulose WP-300 (a protective colloid), 2.00 g of Tergitol NP-40, 2.60 g of Tergitol NP-15 (two nonionic surfactants), 2.20 g of Alcolac Siponate DS-4 (an anionic surfactant), an 0.40 g of ammonium bicarbonate.
3. Blanket the stirring solution with nitrogen and warm the reaction substrate to 55°C for 20 min.
4. Add the monomer seed solution consisting of 12 g of vinyl acetate, 3.00 g of butyl acrylate, and 5.00 g of vinyl neodecanoate followed by 0.16 g of ammonium persulfate.
5. Raise and maintain the temperature to 75°C for 15 min to form the seed latex. (Note any change in the appearance of the reacting mixture.)
6. Place the remaining monomer solution consisting of 108.0 g of vinyl acetate, 27.6 g of butyl acrylate, and 44.2 g of vinyl neodecanoate in the 250-ml addition funnel and a solution of 0.4 g of ammonium persulfate in 40 ml of deionized water in the 50-ml addition funnel.
7. After raising the temperature in the reaction flask to 78°C, begin adding the monomer solution at such a rate that all of it is added reasonably uniformly in 120 min. At the same time, start the addition of the initiator solution at such a rate that all of it is added reasonably uniformly

in 150 min. During this addition period maintain the reaction temperature at 78 ± 2°C.
8. After the additions have been completed, continue heating and stirring for another hour.
9. Cool the latex and filter it through a 200 mesh stainless-steel screen.

REPORT

1. Estimate the percentage of coagulum that has been retained on the filter screen.
2. Determine the percentage of solids of the latex.
3. Calculate the percent yield.

REFERENCE

1. Procedure is based on "Vynate Copolymer Latex For Exterior Architectural Coatings," Technical Bulletin F-60831, Union Carbide Chemicals and Plastics Company, Inc., Danbury, CT (1994), p. 2.

EXPERIMENT 12

Preparation of Poly(vinyl alcohol) by the Alcoholysis of Poly(vinyl acetate)

INTRODUCTION

For all practical purposes, monomeric vinyl alcohol exists only in its tautomeric form as acetaldehyde. Therefore, poly(vinyl alcohol) (PVAlc) is prepared by the hydrolysis of polymers of vinyl esters. For practical reasons, the starting material of choice is poly(vinyl acetate) (PVAc). Although hydrolysis may be carried out under acidic conditions, alkaline conditions in the presence of an alcohol are preferred. The reaction may be represented by

$$-(CH_2CH[OCOCH_3])_n- + nCH_3OH \longrightarrow -(CH_2CH[OH])_n- + nCH_3OCOCH_3$$

It should be noted that methyl acetate is a coproduct of this process. In industrial processes, this product is isolated by distillation as a methanol–methyl acetate azeotrope.

The base-catalyzed alcoholysis of poly(vinyl acetate) is quite rapid and is thought to be autocatalytic. Under usual reaction conditions, the reaction goes to approximately 90% completion. To reach 100% conversion requires specialized conditions. Achieving partial hydrolysis is difficult because of the rapidity of the process.

BACKGROUND

Commercial PVAlc is available in several degrees of hydrolysis as well as in several molecular weight ranges. The PVAlc with a degree of hydrolysis of approximately 81% is of particular interest. This material appears to be the most water soluble of the commercial products.

It should be noted that dissolving PVAlc in water is, at best, troublesome. To avoid lump formation, it is best to sprinkle the product very slowly into rapidly stirred water. Even then, great care and patience are needed. For viscosity measurements in an aqueous solution, concentrations of 4% polymer in water are the norm.

The rate and/or the degree of hydrolysis of PVAc is independent of the molecular weight of the starting material.

However, predicting the molecular weight of the hydrolysis product is another matter. Most commercial PVAc is somewhat cross-linked. The cross-links may involve the methylene group of the vinyl unit or the methyl of the acetate moiety. Base hydrolysis of the former type of cross-link does not affect the chain length of the polymer. The hydrolysis of an ester linkage whose methyl group is attached to another polymeric group will, however, result in the cleavage of the cross-linked chain. The result is a reduction of the average molecular weight of the PVAlc that is being formed. Upon reacetylation of PVAlc with acetic anhydride, a PVA with a lower molecular weight forms. This product is, presumably, not cross-linked. On alcoholysis of this second product, a PVAlc forms that has substantially the same molecular weight as the initially isolated PVAlc [1–4].

A sample of poly(vinyl alcohol) that had been freed of cross-links through the ester portions of the starting material lend themselves to an evaluation of the ratio of head-to-tail chains to head-to-head chains. In the latter case, there should be gem diol structures corresponding to each head-to-head unit. Periodic acid is a specific reagent for the cleavage of gem diols. If this reagent is applied to an aqueous solution of poly(vinyl alcohol), chain scission would take place at the head-to-head moieties. This would result in a reduction in the overall molecular weight of the polymer. By measuring the viscosity molecular weight of the PVAlc before and after treatment with periodic acid, the ratio of the two structural features may be calculated [5].

APPLICABILITY

Poly(vinyl alcohol) is used in films, coatings, fibers, and as a viscosity modifier of a variety of aqueous systems such as certain cosmetics. Its films, formulated with dichromate salts, may be cross-linked by exposure to ultraviolet light. This property has found application in photoengraving and related fields [6].

In the preparation described here, the alkaline reagent of choice is potassium hydroxide because it is more soluble in methanol than is sodium hydroxide. In the final purification of the product, residues of potassium salts and potassium hydroxide are also more easily removed than sodium analogs. Commercially, sodium methoxide is often used as the catalyst for hydrolysis, but handling of this reagent calls for a somewhat more complex procedure.

As small amounts of water accelerate the alcoholysis, it may not be necessary to operate in moisture-free circumstances unless sodium methoxide is used or careful reaction rate studies are undertaken.

SAFETY PRECAUTIONS

Before this experiment is carried out, the student must read the material safety data sheets (MSDS) for all the chemicals used as well as for the products. The instructor must approve that you have read and understood the MSDS for the safe handling of these materials.

Please be advised that all chemicals should be considered hazardous and should be handled in a hood and with proper personal protective equipment (lab coat, proper gloves, approved safety glasses, and/or goggles). Avoid inhaling vapors and/or aerosolized materials. Avoid skin/eye contact with all chemicals at all times. Wash hands frequently. See the instructor if you have any questions or concerns.

APPARATUS

1. 250-ml Erlenmeyer flask with stopper
2. 1-liter three-necked flask

3. Heating mantle
4. Reflux condenser
5. 125-ml addition funnel
6. Mechanical stirrer with appropriate bearing
7. Buchner funnel and filter flask
8. Heating mantle
9. Vacuum desiccator or vacuum oven
10. Mortar and pestle
11. Rubber gloves, face shield, and goggles
12. Miscellaneous beakers, flasks, stirring rods, filter paper, funnels, etc.

REAGENTS AND MATERIALS

1. 10 g poly(vinyl acetate) (molecular weight optional); a medium molecular weight would be Aldrich 18948-0
2. 1 liter of anhydrous methanol
3. 0.5 g anhydrous potassium hydroxide
4. deionized water
5. pH paper

PROCEDURE

1. In a mortar, cautiously and rapidly grind 0.5 g of anhydrous potassium hydroxide to a fine powder with a pestle, transfer the powder to a 250-ml Erlenmeyer flask, and add 100 ml methanol. Stopper the flask and disperse the potassium hydroxide. Preserve the material until required.
2. In a 250-ml three-necked flask fitted with a mechanical stirrer and a reflux condenser, place 10 g of poly(vinyl acetate) and 100 ml of anhydrous methanol. Close the flask with an addition funnel, turn on the condenser, and, with vigorous stirring, heat the mixture at reflux until the polymer has dissolved. Reduce the heat to a moderate rate of reflux and add 20 ml of the solution of potassium hydroxide in methanol drop by drop through the addition funnel. The reaction rate may be moderated, if necessary, by adding more methanol or by lowering the heating mantle and cooling so as to keep the temperature at below 60°C.
3. After the reaction has slowed down, continue heating at reflux for 2 hr (see Note).
4. Cool the flask. Remove stopcock grease from all the joints of the flask. With suction, filter off the poly(vinyl alcohol) that has formed. Wash the product repeatedly with methanol until a sample of the methanol washings, diluted with deionized water, have a pH below 7.5 (i.e., until residual alkali has been removed).
5. Collect the product on a Buchner funnel and draw air through the product until visually dry. The final drying of the product to constant weight may be carried out in a vacuum desiccator or in a vacuum oven.

NOTE

Because PVAlc is insoluble in methanol, it usually precipitates as it forms. Occasionally, a gel forms instead of a powder. After cooling, such a gel may be broken up in a Waring blender with additional methanol and worked up as usual.

REFERENCES

1. W. H. McDowell and W. O. Kenyon, *J. Am. Chem. Soc.* **62,** 415 (1940).
2. L. M. Minsk, W. J. Priest, and W. O. Kenyon, *J. Am. Chem. Soc.* **63,** 2715 (1941).
3. O. L. Wheeler, S. L. Ernst, and R. H. Crozier, *J. Polym. Sci.* **8,** 409 (1952).
4. S. R. Sandler and W. Karo, "Polymer Syntheses," 2nd Ed., Vol. III, Academic Press, San Diego, 1996.
5. P. J. Flory and F. S. Leutner, *J. Polym. Sci.* **3,** 880 (1948).
6. D. L. Cincera, in "Kirk-Othmer, Encycl. of Chem. Tech.," 3rd Ed., p. 864, Wiley, New York, 1983.

SECTION II

Polymer Characterization

CHAPTER 13

Nuclear Magnetic Resonance

INTRODUCTION

Nuclear magnetic resonance (NMR) spectroscopy is a most effective and significant method for observing the structure and dynamics of polymer chains both in solution and in the solid state [1]. Undoubtedly the widest application of NMR spectroscopy is in the field of structure determination. The identification of certain atoms or groups in a molecule as well as their position relative to each other can be obtained by one-, two-, and three-dimensional NMR. Of importance to polymerization of vinyl monomers is the orientation of each vinyl monomer unit to the growing chain tacticity. The time scale involved in NMR measurements makes it possible to study certain rate processes, including chemical reaction rates. Other applications are isomerism, internal relaxation, conformational analysis, and tautomerism.

PRINCIPLES

Nuclear magnetic resonance spectroscopy is a form of absorption spectroscopy and concerns radio frequency (rf)-induced transitions between quantized energy states of nuclei that have been oriented by magnetic fields. Several nonmathematical introductions to NMR are recommended to supplement the material here [1–9]. For greater mathematical depth, a number of excellent texts are available [10–26].

Theory

All nuclei carry a charge. In some nuclei, this charge "spins" on the nuclear axis. This circulation of nuclear charge generates a magnetic dipole along the axis. The intrinsic magnitude of the generated dipole is expressed in terms of the nuclear magnetic moment, μ. The angular momentum of the spinning charge can be described in terms of spin quantum numbers, **I**, with values 0, $1/2$, 1, $3/2$, 2, etc. (**I** = 0 denotes a nucleus with no spin.)

Each proton and neutron has a spin. **I** is the resultant of these spins. If both of the protons and neutrons are even, **I** is zero and there is no NMR signal (e.g., ^{12}C, ^{16}O). Those nuclei having a nuclear spin of $1/2$ (e.g., ^{1}H, ^{13}C, ^{31}P, ^{19}F, ^{119}Sn, ^{29}Si, ^{15}N) have a uniform spherical charge distribution. Nuclei having nuclear spins greater than $1/2$ (e.g., ^{14}N, ^{17}O, ^{7}Li, ^{35}Cl) have a nonspherical charge distribution that gives rise to a nuclear quadrupole moment. Only nuclei with $I = 1/2$ will be considered.

The spin quantum number **I** determines the number of orientation a nucleus may assume in an external uniform magnetic field in accordance with the formula $2I + 1$. For a spin $I = 1/2$ nuclei, there are two orientations in an applied uniform magnetic field: parallel with the applied magnetic field (aligned with the field) or antiparallel (aligned against the field). The former is the low energy state, whereas the latter is the high energy state (see Fig. 13.1). Energy levels are a function of the magnitude of the nuclear magnetic moment, μ, and the strength of the applied external field, B_0. The energy difference between the upper and the lower energy states is in the radio frequency range. Each magnetic nucleus will absorb a specific radio frequency and "flip" from the lower energy state to the upper energy state. After absorption of energy by the nuclei, they must have a mechanism whereby they can dissipate the energy and return to the lower energy state. There are two primary processes for relaxation to the lower energy state: (a) spin–lattice or longitudinal relaxation and (b) spin–spin or transverse relaxation. Spin–lattice relaxation, T_1, involves the transfer of energy from the nuclei to the environment of the molecular lattice. Spin–spin relaxation, T_2, arises from direct interactions between spins of different nuclei that can cause transverse relaxation without any energy transfer to the lattice. The nucleus is shielded by its electron cloud. The unique aspect of NMR is that nuclei in different parts of the molecule will have different electronic environments depending on the nature of the bonding and other surrounding groups, which leads to differences in the field required to realign or "flip" the spin of the nuclei and consequently to different absorption frequencies. This is referred to as the chemical shift from a standard reference. Reference materials are materials that are chemically inert, magnetically isotropic, and dissolve in the same solvent as the compound to be analyzed. Tetramethylsilane (TMS) is used as a chemical shift reference for proton, carbon-13, and silicon-29 analysis in organic solvents and is given 0 as its chemical shift position. The difference in absorption or shift from TMS or another reference is usually expressed in dimensionless units of parts per million, which is the shift in cycles (Hz) from TMS divided by the applied frequency of the instrument and multiplying by a factor of 10^6, i.e.,

$$\delta = \frac{\text{frequency of absorption}}{\text{applied frequency}} \times 10^6.$$

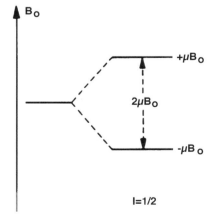

Fig. 13.1 A nucleus with spin $1/2$ in an applied magnetic field.

Because δ units are expressed in parts per million, the expression ppm is used. Thus a peak at 400 Hz from TMS at an applied frequency of 400 MHz would be at δ 1.00 or 1.00 ppm. NMR spectrometers with field strengths as high as 17.6 Tesla (T) or 750 MHz are commercially available.

The NMR spectrum provides detailed information for polymer characterization. Some significant parameters are frequency or chemical shift, intensity (or peak area), line width, J-coupling constants, and relaxation rates (T_1 and T_2). The NMR spectrum is a series of absorption peaks representing nuclei in different chemical/electronic environments (Fig. 13.2). Each absorption area is proportional to the number of nuclei it represents, which provides considerable information about the molecule. However, there is another complication or refinement to structure determination that involves spin–spin coupling, an indirect coupling of the nuclei spins through the intervening bonding electrons. Briefly, it occurs because there is some tendency for a bonding electron to pair its spin with the spin of the nearest spin $1/2$ nuclei; the spin of a bonding electron, having been thus influenced, the electron will affect the spin of the other bonding electron and so on through to the next spin $1/2$ nuclei. Coupling for protons is ordinarily not important beyond three bonds unless there is a bond delocalization as in unsaturated or aromatic systems or ring strain as in small rings or bridged systems.

Suppose two protons are in very different chemical environments from one another, as in an absorbing proton (three chemically identical protons of a methyl group) on a carbon atom, which has as a nearest neighbor a carbon having one proton on it, designated as H_α.

$$H_\alpha - \underset{|}{\overset{|}{C}} - CH_3$$

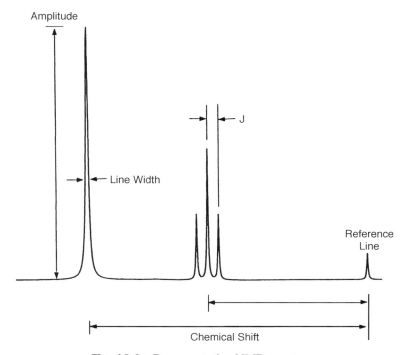

Fig. 13.2 Representative NMR spectrum.

Each proton will give rise to an absorption; the areas will be one (methine) to three (methyl). The absorptions will be separated. However, the spin of each type of proton is affected by the two orientations of the other proton through the intervening bonding electrons such that a splitting of the absorption signals will be observed. The splitting patterns can usually be interpreted by using two rules: (1) the multiplicity of the splitting is determined by $2n\mathbf{I} + 1$, where $\mathbf{I}$ is the spin quantum number and n is the number of nuclei giving rise to the splitting and (2) the multiplicity and relative intensities may be easily obtained from Pascal's triangle $(a + b)^n$.

Polymer Characterization

Polymer Chemistry

In general, there are two distinctively different classes of polymerization: (a) addition or chain growth polymerization and (b) condensation or step growth polymerization. In the former, the polymers are synthesized by the addition of one unsaturated unit to another, resulting in the loss of multiple bonds. Some examples of addition polymers are (a) poly(ethylene), (b) poly(vinyl chloride), (c) poly(methyl methacrylate), and (d) poly(butadiene). The polymerization is initiated by a free radical, which is generated from one of several easily decomposed compounds. Examples of free radical initiators include (a) benzoyl peroxide, (b) di-tert-butyl peroxide, and (c) azobiisobutyronitrile.

Other macromolecules are formed by condensing their monomers to form a repeat functional group (e.g., esters, amides, ethers) interspersed by alkyl chains, aromatic rings, or combinations of both. These condensations are characterized frequently, although not always by the loss of some by product (e.g., water, alcohol). The methods of formation of these polymers are far more varied than those of addition polymers. Examples of condensation polymers are (a) poly(esters), (b) poly(urethanes), (c) poly(carbonate), and (d) poly(phenylene oxide).

Polymer Tacticity

After the polymers are prepared, several other considerations arise, such as molecular weight [26] and crystallinity. Another major consideration, especially for asymmetric vinyl monomers, is the orientation of each monomer adding to the growing chain, called *tacticity*. There are three types of tacticity: isotactic, syndiotactic, and atactic [27]. If all the monomeric units possess the same enantiomorphic configuration [28], it is termed *isotactic*. In isotactic polymers, all of the R-substituents appear on the same side of the polymer backbone. If the enantiomorphic configuration along the polymer chain alternates regularly, i.e., the R-substituents alternate regularly from one side of the polymer chain to the other along the polymer backbone, the polymer configuration is *syndiotactic*. If there is no long range order to the placement of the R-substituents of the monomers, the polymer is said to be *atactic*. Tacticity is demonstrated in Fig. 13.3.

In polymers that exhibit tacticity, the extent of the stereoregularity determines the crystallinity and the physical properties of the polymers. The placement of the monomer units in the polymer is controlled first by the steric and electronic characteristics of the monomer. However, the presence or absence of tacticity, as well as the type of tacticity, is controlled by the catalyst employed in the polymerization reaction. Some common polymers, which can be prepared in specific configuration, include poly(olefins), poly(styrene), poly(methyl methacrylate), and poly(butadiene).

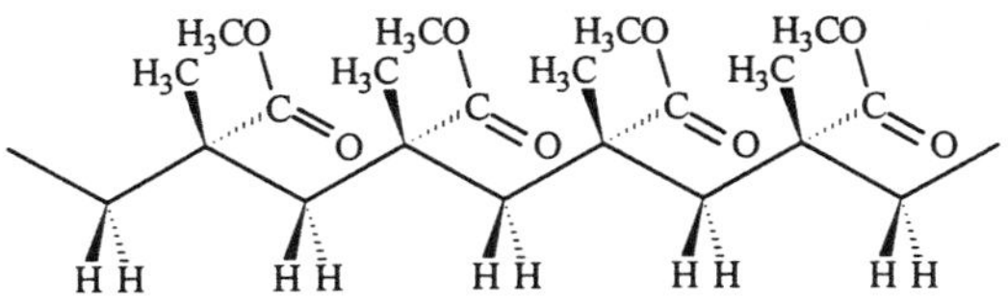

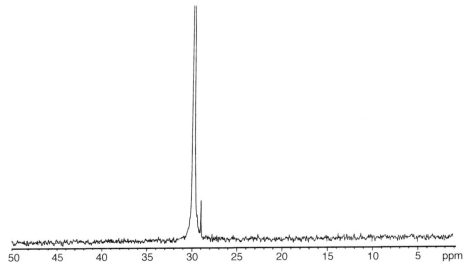

isotactic PMMA

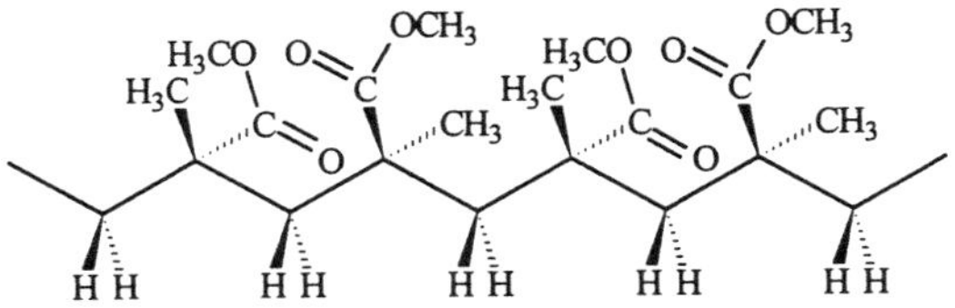

syndiotactic PMMA

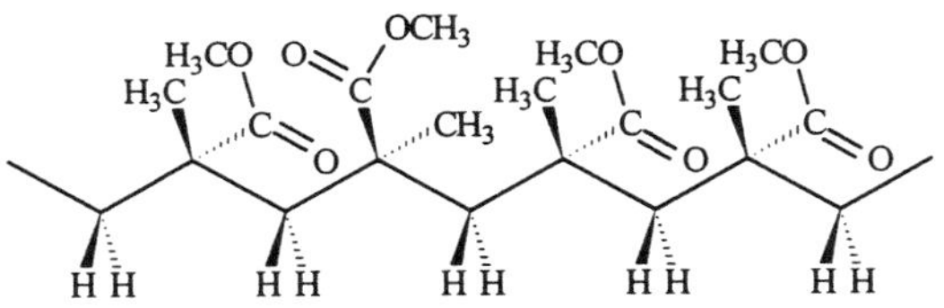

atactic PMMA

Fig. 13.3 Illustration of tacticity.

Homopolymer Characterization

NMR spectroscopy is sensitive to the polymer composition and stereochemistry, branching, isomerism, head-to-head and tail-to-tail additions, and chain ends. Figures 13.4 and 13.5 show two types of poly(ethylene) [29]. Figure 13.4 is Chevron HDPE LX-1159, which is a high-density poly(ethylene), whereas Fig.

Fig. 13.4 Proton-decoupled, 75.4-MHz carbon-13 NMR spectrum of Chevron HDPE LX-1159, a high-density poly(ethylene).

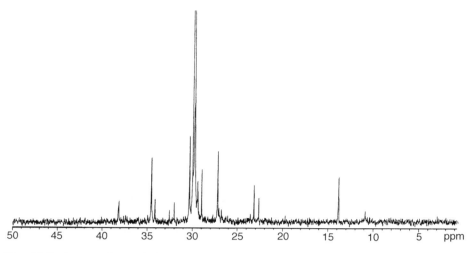

Fig. 13.5 Proton-decoupled, 75.4-MHZ carbon-13 NMR spectrum of Chevron LDPE 5422, a low-density poly(ethylene).

13.5 is Chevron LDPE 5422, a low-density poly(ethylene). The number of branches per 1000 carbon atoms may be calculated (Table 13.1) [26,27].

It is possible to calculate the number average molecular weight, M_n, using carbon-13 NMR (Table 13.2) [26,29,30]. These data compare well with those obtained from gel permeation chromatography (GPC).

Spectral Editing and Two-Dimensional NMR

Both solution-state and solid-state NMR spectroscopy are important analytical tools used to study the structure and dynamics of polymers. This analysis is often limited by peak overlap, which can prevent accurate signal assignment of the dipolar and scalar couplings used to determine structure/property relationships in polymers. Consequently, spectral editing techniques and two- or more dimensional techniques were developed to minimize the effect of spectral overlap. This section highlights only a few of the possible experiments that could be performed to determine the structure of a polymer.

Figure 13.6 is the proton-decoupled carbon-13 NMR distortionless enhancement of polarization transfer (DEPT) spectra of poly(methyl-1-pentene) [29]. This experiment, after data manipulation, separates the methine, methylene, and

TABLE 13.1
Calculating the Number of Branches

Sample description	Branches per 1000 carbons	
	Branch type	Amount
Chevron HDPE LX-1159	Not detected	—
Chevron LDPE 5422	Butyl	8.1
	Amyl	2.1
	Hexyl or greater	5.7

TABLE 13.2
Calculating Molecular Weight

Sample description	Molecular weight (M_n)	
	NMR[29]	GPC[30]
Chevron HDPE LX-1159	25,900	22,200
Tonen TR-009	16,600	15,300

methyl carbons into separate plots. The DEPT experiment does not permit the observation of quaternary carbons. If it is desirable to determine the frequency position of the quaternary carbons, then an attached proton test (APT) could be performed.

Figures 13.7 and 13.8 are two examples of two-dimensional NMR spectroscopy applied to polymers. Figure 13.7 is the proton homonuclear correlated spectroscopy (COSY) contour plot of Allied 8207A poly(amide) 6 [29]. In this experiment, the "normal" NMR spectrum is along the diagonal. Whenever a "cross peak" occurs, it is indicative of protons that are three bonds apart. Consequently, the backbone methylenes of this particular polymer can be traced through their J-coupling. Figure 13.8 is the proton–carbon correlated (HETCOR) contour plot of Nylon 6 [29]. This experiment permits the "mapping" of the proton resonances into the carbon-13 resonances.

Solid-State NMR

The commercial value of high-performance polymers, i.e., polymers of high molecular weight, which may be composite structures and do not dissolve, have driven the development of new methods for material characterization. Solid-state NMR spectra can be acquired in either high resolution or wide-line mode, methods that yield complementary information [8,13,19,23]. Solid-state NMR has the advantage of being able to analyze the material "as received"; consequently, it

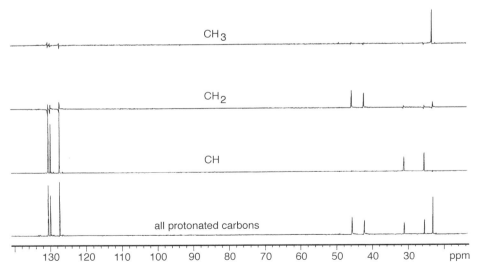

Fig. 13.6 Proton-decoupled, 75.4-MHz carbon-13 NMR distortionless enhancement of polarization transfer (DEPT) spectra of poly(methyl-1-pentene) in 1,2,4-trichlorobenzene.

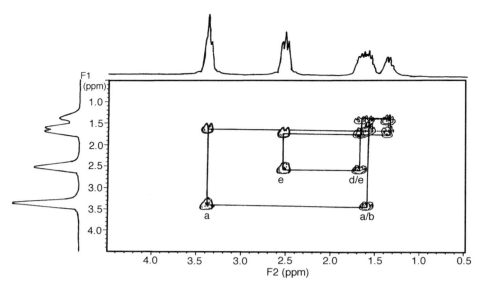

Fig. 13.7 Proton homonuclear correlated spectroscopy (COSY) contour plot of Allied 8207A, a poly(amide) 6.

bridges the gap between traditional solution techniques (titration, solution-state NMR, etc.) and physical property measurements (thermal analysis, x-ray, etc.). Solid-state NMR can provide information not only about the polymer composition, but also about its crystallinity. Figure 13.9 shows comparison spectra of poly(ethylene) under several solid-state NMR data acquisition conditions. Figure 13.9a shows the cross polarization magic angle spinning (CP/MAS) carbon-13; Fig. 13.9b demonstrates the progressive saturation experiment, which filters out

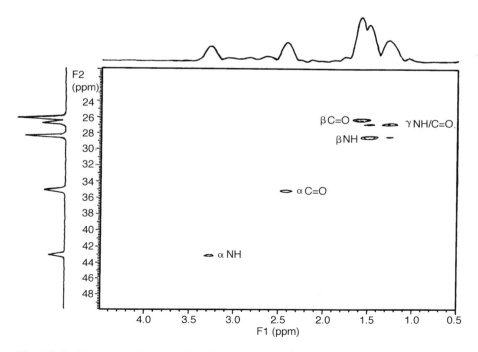

Fig. 13.8 Proton–carbon correlated spectroscopy (HETCOR) contour plot of nylon 6.

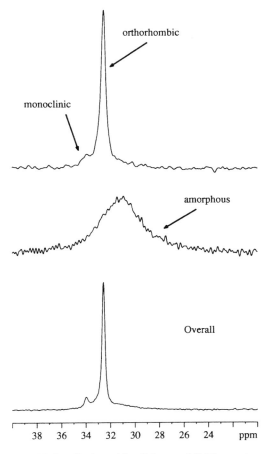

Fig. 13.9 Carbon-13 solid-state NMR spectra of poly(ethylene): (a) cross-polarization magic angle spinning (CP/MAS), (b) progressive saturation, and (c) Torchia T_1 experiment.

long T_1 components out of the spectrum, and Fig. 13.9c shows the Torchia T_1 experiment in which components with a short T_1 are filtered out of the experiment.

ACCURACY AND PRECISION

In general, the area under an absorption band is proportional to the number of nuclei responsible for the absorption. The integral is represented as a sigmoidal function; the height of each integral is proportional to the number of nuclei in that particular region of the spectrum. Accuracy of reproducing the integral should be within ± 2%. Exact phase correction of the Fourier transformed signal is crucial to accurately reproducing the integral. Whenever the empirical formula is known, the total height of the integral (in any arbitrary unit) divided by the number of nuclei responsible for the absorption yields the increment of height per nucleus. This is related to the number of moles of that nuclei or compound. Overlapping bands can be resolved either by using peak deconvolution programs available on the spectrometer or off-line on a personal computer or by a more traditional "cut and weigh."

SAFETY PRECAUTIONS

Safety glasses must be worn in the laboratory at all times. Appropriate safety gloves are recommended when making up samples.

Although there is no known health risks of magnetic fields, care must be taken when working within the fringe field of the magnet as metal objects can be "captured" by the magnet (usually 3 to 10 feet from the center of a superconductive magnet; actual distance depends on the magnetic field strength. Refer to the vendor's manual for specific distance).

Material safety data sheets (MSDS) for all chemicals being used must be read prior to beginning the experiment. All chemicals should be considered hazardous from a standpoint of flammability and toxicity.

APPARATUS

1. A NMR spectrometer with the following capabilities: (a) multinuclear observe frequencies, (b) proton decoupling, and (c) variable temperature
2. Clean, high-resolution NMR tubes; quality of the tube is determined by the field strength of the magnet (5 mm OD for proton, 10 mm OD for carbon-13)
3. Balance
4. Test tubes
5. Disposable pipettes
6. Cotton balls
7. Heating block

REAGENTS AND MATERIALS

Chemicals for Epoxide Kinetics

1. Poly(Bisphenol A-co-epichlorohydrin)-glycidyl end capped-CAS[25036-25-3], ave. mol. wt. 348, m.p. 41–44°C; d 1.169.
2. Diethylenetriamine, 99% CAS[111-40-0], $H_2NCH_2CH_2NHCH_2CH_2NH_2$, mol. wt. 103.17, m.p. −35°C, d 0.955.
3. Triethylenetetramine, 60% CAS[112-24-3], $H_2NCH_2CH_2NHCH_2CH_2NHCH_2CH_2NH_2$, mol. wt. 146.24, m.p. 12°C, d 0.982.
4. Deuterated dimethyl sulfoxide (DMSO-d_6), >99%, isotopically pure.
5. Tetramethylsilane

Chemicals for Tacticity

1. Poly(methyl methacrylate) samples of different tacticities.
2. Deuterated chloroform, $CDCl_3$, >99%, isotopically pure.
3. 1,2,4-Trichlorobenzene, spectroscopy grade.
4. Carbon tetrachloride, spectroscopy grade.

PROCEDURE

The operating instructions for the NMR spectrometer should be carefully read and understood before any experimental work is attempted. Recommended preliminary reading includes Refs. 1–16 as well as ASTM E 386-90.

Epoxy Kinetics: Condensation Reaction

Following is the polymerization procedure of a cured epoxy resin, Bisphenol-A diglycidylether, with polyamines. Record the time of this experiment. *Note:* Some NMR spectrometers may have kinetics experiments as macros.

Preparation of the Reference Solutions

a. Weigh out 40 mg of Bisphenol-A diglycidyl ether and place in a high-quality, high-resolution NMR tube.
b. Add 0.8 ml of DMSO-d_6 to the NMR tube containing the Bisphenol-A diglycidyl ether.
c. Add a small drop of TMS as a chemical shift reference.
d. Prepare similar solutions of amines.
e. Acquire the proton spectra of these reference solutions. Acquire 16 to 32 transients, with a delay time between acquisitions of 30 sec.
f. Set the chemical shift position of TMS to 0.0 ppm (see instrument manual for the exact procedure).
g. Record all the chemical shift positions of the ether and amine resonances.
h. If time permits, acquire the proton-decoupled carbon-13 NMR spectra, both in "broad band" (qualitative) and suppressed Overhauser (quantitative) mode. (See instrument manual for the exact procedure.)
i. Weigh out 10.44 g (0.03 mol or 0.06 equivalent) of the Bisphenol A epoxy resin into a test tube.
j. Weigh out 2.06 g (0.02 mol or 0.06 equivalent) of diethylenetriamine into a test tube.
k. Note the time; time = 0 is when the chemicals are mixed.
l. Add the diethylenetriamine to the test tube containing Bisphenol A epoxy resin.
m. Mix well with a spatula.
n. Quickly take a small amount of the reaction mixture (~40 mg) and place in an NMR tube.
o. Add 0.8 ml of DMSO-d_6 and mix well.
p. Place the NMR tube into the spectrometer.
q. Record time again as data begin to be recorded. This is time = t_1.
r. Take the proton NMR spectra at 15-min intervals and store the data acquisition file for the next 1.5 hr.
s. Plot NMR spectra for each of the time intervals.
t. Integrate the peaks.
u. Make a plot of the intensity of the epoxide peak vs time and the amine peak vs time to follow the rate of reaction.
v. Store the samples until the next laboratory period. If the samples are still soluble, acquire, store, and plot the proton spectrum. Record any difference.

Polymer Tacticity

The preparation of the polymer samples should require at least one more lab, and directions in the literature cover this adequately.

The three types of poly(methyl methacrylate) show different physical properties as seen in Table 13.3.

The proton and carbon-13 NMR absorptions for each configuration are given in Table 13.4.

TABLE 13.3
Different Physical Properties of Poly(methyl methacrylate)

Tacticity	Glass transition (Tg/°C)	Density
Isotactic	45	1.22
Syndiotactic	115	1.19
Atactic	104	1.19

a. Prepare the NMR samples in an appropriate solvent depending on the nucleus to be studied (see Notes 1–3).
b. Carefully put the NMR tube into the spinner, being careful to follow the probe depth gauge instructions. Place the spinner into the magnet and spin.
c. Check to be sure that the NMR probe is at the appropriate temperature. Allow the sample 10 min to thermally equilibrate within the probe before attempting to shim the magnet field to achieve the best homogeneity. Once the sample resolution is at the desired level, begin data acquisition.
d. For proton NMR, set the operating parameters such that the sweep width encompasses −2 to 12 ppm and adjust the digital resolution so that there is at least a 2- to 3-Hz/data point; a 90° pulse is used with a delay between pulses of 15 sec. Acquire 16 to 64 transients depending on the signal-to-noise ratio. Set the methyl peak of TMS to 0.00 ppm.
e. Integrate the peaks of the PMMA. (For NMR spectrometers, whose proton frequency is 60 MHz or greater, the backbone methyl groups will be well separated and can be integrated easily.)

TABLE 13.4
Proton and Carbon-13 NMR Absorptions

	Tacticity					
	Isotactic		Syndiotactic		Atactic	
Polymer group	^{1}H	^{13}C	^{1}H	^{13}C	^{1}H	^{13}C
Ester methyl O—CH$_3$	Singlet 3.5	51	Singlet 3.5	51	Singlet 3.5	51
Ester carbonyl C=O	—	22.0	—	18.5	—	20.5
Quaternary carbon —C—	—	47	—	45.5	—	46
Methylene —CH$_2$—	Quartet 1.4, 1.6, 2.1, 2.3	Multiplet 54–57	Singlet 1.86	Multiplet 54–57	Multiplet 54–57	Multiplet 54–57
Backbone methyl —CH$_3$	1.33	Multiplet 176	1, 12	Multiplet 178	1.23	Multiplet 177

CALCULATIONS

Epoxy Kinetics

a. Calculate the mole percent of each monomer from the integrated areas of the proton NMR spectrum for all acquisition times.
b. Calculate the weight percent of each monomer from the mole percentage determined above.
c. How does the proton NMR spectrum of the reaction solution change as a function of time?
d. Speculate on the utility of this type of experiment.

Polymer Tacticity

a. Calculate the mole percent of each type of tacticity present in the polymer solutions from the backbone methyl peaks proton NMR spectra.
b. Calculate the mole percents of each type of tacticity using the methylene carbons.
c. Compare the mole percents of the unknown polymer to those of the standards that have been run.

REPORT

1. Describe the instrument and experiment in your own words.
2. Include all data obtained.
3. Tabulate kinetic data or polymer tacticity data, depending on the experiment performed.
4. Plot calculated kinetic data to obtain information about the order of the kinetic reaction.

NOTES

1. Samples for proton NMR should be made up as 2 to 8 weight percentage (20–80 mg/1 ml) depending on the magnetic field strength and probe sensitivity of the NMR. Samples for quantitative carbon-13 NMR should be made up as 10–20 weight percentages (200 to 400 mg/2 ml) depending on the polymer molecular weight, as well as the same instrumental conditions for proton. The proton NMR experiments can be conducted using deuterated chloroform, $CDCl_3$, as the solvent and NMR probe at 50°C. The carbon-13 NMR experiments can be conducted using 1,2,4-trichlorobenzene, running the NMR in the "unlocked" condition with the probe at 120+°C.
2. Because the polymer requires at least 30 min to dissolve, even using a heating block, it would be preferable to prepare the solutions at least a day in advance.
3. The polymer solutions can be made directly in the NMR tubes unless it is believed that the polymer contains gel and that this gel will be a problem.

    a. If the polymer contains gel, the polymer solutions can be made up in test tubes. These samples will be filtered before being put in the NMR tube. The following small filter assembly is made from a disposable

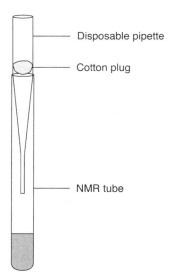

Fig. 13.10 Filtering the polymer into the NMR tube.

pipette and cotton ball. A filter assembly must be made for each polymer solution. Take a small amount of cotton (angel hair), place it in the upper end of a disposable pipette, and push it in toward the top (see Fig. 13.10). Place the pipette in a test tube. Rinse the cotton ball on the pipette with 1–2 ml carbon tetrachloride to remove loose fibers and allow the assembly to drain completely into the test tube so as to not dilute the sample. Next, place the top of the assembly into the NMR tube. After the polymer solution has drained into the NMR tube, remove the assembly. Add a small drop of TMS to the polymer solution. Carefully put the cap into the NMR tube. Discard the filter assembly appropriately.

ACKNOWLEDGMENTS

The authors thank Dr. Russell Lewis, Elf Atochem North America, for proofreading this manuscript and providing corrections and figures. The authors also thank Mr. Tuan-Dung Nguyen, Elf Atochem North America, for helping to procure the samples and for running some of the NMR spectra.

REFERENCES

1. "Comprehensive Polymer Science: The Synthesis, Characterization Reactions and Applications of Polymers" (G. Allen and J. C. Bevington, eds.), Vol. 1, Chaps. 17, 18, and 19 and references therein, Pergamon Press, New York, 1989.
2. F. A. Bovey, "NMR Spectroscopy," Academic Press, New York, 1969.
3. R. H. Bible, Jr., "Interpretation of NMR Spectra," Plenum Press, New York, 1965.
4. R. M. Silverstein, G. C. Bassler, and T. C. Morrill, "Spectrometric Identification of Organic Compounds," 3rd Ed. Wiley, New York, 1974.
5. T. C. Farrar and E. D. Becker, "Pulse and Fourier Transform NMR," Academic Press, New York, 1971.
6. K. Mullen and P. S. Pregasen, "Fourier Transform NMR Techniques: A Practical Approach," Academic Press, London, 1976.
7. L. M. Jackman and S. Sternhill, "Applications of NMR Spectroscopy in Organic Chemistry," 2nd Ed., Pergamon Press, New York, 1969.

8. P. A. Mirau, "Polymer Characterization" (B. J. Hunt and M. I. James, eds.), Chap. 3, Blackie Academic and Professional, New York, 1993.
9. F. A. Bovey, "Chain Structures and Conformations of Macromolecules," Academic Press, New York, 1982.
10. A. Abragam, "The Principles of Nuclear Magnetism," Oxford University Press, Oxford, 1961.
11. A. Carrington and A. D. McLachlan, "Introduction to Magnetic Resonance: With Applications to Chemistry and Chemical Physics," Harper and Row, New York, 1967.
12. C. P. Slichter, "Principles of Magnetic Resonance," Springer Verlag, Berlin, 1978.
13. G. E. Martin and A. S. Zektzer, "Two-Dimensional NMR Methods for Establishing Connectivity," VCH Publishers, New York, 1988.
14. H. Günther, "NMR Spectroscopy: An Introduction," Wiley, New York, 1980.
15. C. A. Fyfe, "Solid State NMR for Chemists," C.F.C. Press, Ontario, Canada, 1983.
16. M. L. Martin, J.-J. Delpuech, and G. J. Martin, "Practical NMR Spectroscopy," Heyden and Sons, London, 1980.
17. R. K. Harris, "Nuclear Magnetic Resonance Spectroscopy: A Physiochemistry View," Pitman, London, 1983.
18. J. A. Pople, W. G. Schneider, and H. J. Bernstein, "High Resolution Nuclear Magnetic Resonance," McGraw-Hill, New York, 1959.
19. A. E. Derome, "Modern NMR Techniques for Chemistry Research," Pergamon Press, 1987.
20. F. A. Bovey, "High Resolution NMR of Macromolecules," Academic Press, New York, 1972.
21. "Solid State NMR of Polymers" (L. J. Mathias, ed.), Plenum Press, New York, 1991.
22. A. E. Tonelli, "NMR Spectroscopy and Polymer Microstructure: The Conformational Connection," VCH Publishers, New York, 1989.
23. J. C. Randall, Jr., "Polymer Sequence Determination: Carbon-13 NMR Method," Academic Press, New York, 1977.
24. "NMR and Macromolecules" (J. C. Randall, Jr., ed.), ACS Symposium Series, No. 247, American Chemical Society, 1984.
25. "High Resolution NMR Spectroscopy of Synthetic Polymers in Bulk," (R. A. Komoroski, ed.), VCH Publisher, Deerfield Beach, Florida, 1986.
26. J. C. Randall, Jr., in "Polymer Characterization by ESR and NMR" (A. E. Woodward and F. A. Bovey, eds.), ACS Symposium Series 142, American Chemical Society, Washington, DC, 1980.
27. Polymer tacticity; These terms are first recorded in *J. Polymer Sci.* **20,** 251, 1956. The words originate from Greek words that prefix tatto (put in order): a) iso- (the same), syndo- (every two and a- (not at all).
28. R. T. Morrison and R. N. Boyd, "Organic Chemistry," 3rd Ed., Allyn and Bacon, Inc., Boston, 1973.
29. J. K. Bonesteel, unpublished work.
30. G. Schroeder, unpublished work.
31. ASTM E 386-90, "Standard Practice for Data Presentation Relating to High-Resolution Nuclear Magnetic Resonance (NMR) Spectroscopy, 1990, reapproved 1995.

EXPERIMENT 14

Infrared Spectroscopy

INTRODUCTION

Infrared radiation was discovered in 1800 by Sir William Herschel [1]. Infrared absorption (IR) investigation of materials began in 1900 [1]. The first commercial IR spectrometers were available at the end of World War II [2]. Studies of polymers were among the first applications of the method [2,3]. Infrared spectroscopy is probably the method most extensively used for the investigation of polymer structure and the analysis of functional groups. [4–26] IR spectrometers have been used to study samples in the gaseous, liquid, and solid state, depending on the types of accessories used. IR has been used to characterize polymer blends, dynamics, surfaces, and interfaces, as well as chromatographic effluents and degradation products. It is capable of qualitative identification of the structure of unknown materials as well as the quantitative measurement of the components in a complex mixture. There are compilations of infrared spectra including indices to collections of spectra and to the literature [27].

BACKGROUND

Electromagnetic Radiation Spectrum

Infrared radiation refers broadly to that part of the electromagnetic radiation between visible and microwave regions. The infrared radiation region is roughly divided into three regions: near-infrared, mid-infrared, and far-infrared (see Table 14.1). In the near-infrared region (NIR), many absorption bands resulting from harmonic overtones and combination bands of the fundamental molecular vibrations are found and are due to the anharmonic nature of these molecular vibrations. The NIR region is dominated by overtones and combination bands of the stretching frequencies associated with C—H, O—H, and N—H because of the high frequency of these vibrations and of the large deviation of the light hydrogen atom from harmonic behavior.

The mid-infrared region is divided into the "group frequency" region (4000–1300 cm^{-1} or 2.5–8 μm) and the "fingerprint" region (1300–650 cm^{-1} or 8.0–15.4 μm). In the group frequency region, the principle absorption bands may be assigned to vibration units consisting of only two atoms of a molecule. These units are more or less dependent only on the functional group giving the absorption and not on the complete molecular structure. Major factors in the

TABLE 14.1
Three Regions of Infrared Radiation

Region	Wave numbers (cm^{-1})	μm	Vibration observed
Near-infrared	14,290–4000	0.7–2.5	Overtones and combination vibration
Mid-infrared	4000–666	2.5–15.0	Fundamental vibration
			Stretching and bending
			Vibrations of light weight atoms
Far-infrared	700–200	14.3–50	Lattice vibrations
			Skeletal vibrations
			Heavy atoms

fingerprint region are single bond stretching frequencies and bending vibrations (skeletal frequencies) of polyatomic systems, which involve motions of bonds linking a substituent group to the remainder of the molecule. The multiplicity is too great for assured individual identification, but collectively the absorption bands aid in identification (Fig. 14.1).

The far-infrared region contains the bending vibrations of carbon, nitrogen, oxygen, and fluorine with atoms heavier than mass 19 and additional bending motions in cyclic or unsaturated systems [28]. The low frequency molecular vibrations are particularly sensitive to changes in the overall structure of the molecule. Far-infrared is particularly useful when studying the conformation of a molecule as the far-infrared bands often differ in a predictable manner for different isomeric forms of the same compound. This region is useful in the

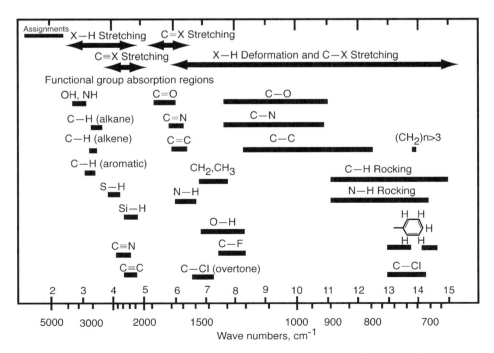

Fig. 14.1 Polymer infrared absorption bands of interest arranged by approximate wavelength and frequency.

study of coordination bonds, particularly organometallic or inorganic compounds whose atoms are heavy and whose bonds are often weak [29].

Theory

Because principles of vibrational spectroscopy have been covered extensively in a number of texts [8,9,28–32], only a few basic points are summarized here.

At temperatures above absolute zero, all atoms in molecules are in continuous vibration with respect to each other [26]. Infrared spectroscopy is an absorption spectroscopy. Two primary conditions must be fulfilled for infrared absorption to occur. First, the energy of the radiation must coincide with the energy difference between excited and ground states of the molecule, i.e., it is quantized (Fig. 14.2). Radiant energy will then be absorbed by the molecule, increasing its natural vibration. Second, the vibration must entail a change in the electrical dipole moment (Fig. 14.3).

The intensity of an infrared absorption band is proportional to the square of the rate of change of dipole moment with respect to the displacement of the atoms. The magnitude of the change in dipole moment may be quite small in some cases, producing weak absorption bands, as seen in the relatively nonpolar C=N group. Conversely, the large permanent dipole moment of the carbonyl group, C=O, causes strong absorption bands. If no dipole moment is created, then no radiation is absorbed and the vibrational mode is said to be infrared inactive. An example of this is the a C=C bond located symmetrically in a molecule.

The major types of molecular vibrations are stretching and bending [26]. IR radiation is absorbed and the associated energy is converted into these types of motions. Consequently, vibrational spectra appear as bands rather than lines because a single vibrational energy change is accompanied by a number of rotational energy changes (Fig. 14.4) [26]. The frequency or wavelength of absorption depends on the relative masses of the atoms, the force constants of the bonds, and the geometry of the atoms in the molecule.

Polymer Characterization

Qualitative Vibrational Analysis (Fingerprinting)

The aim of qualitative analysis of homopolymers by infrared spectroscopy is the elucidation of polymer structure and compound identification. This often entails the identification of the functional groups and the modes of attachment to the polymer backbone [2,4,25,26]. In the case of mixtures, the aim of qualitative

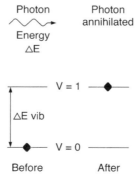

Fig. 14.2 Absorption of infrared radiation.

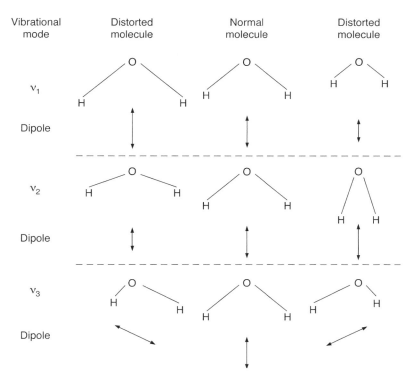

Fig. 14.3 Infrared vibrations of a water molecule.

analysis is to determine the presence of individual components [2,4,25,26]. In the case of copolymers, it is to determine the presence of individual monomer units [2,4,25,26]. The IR spectrum of a mixture is additively composed of spectra of the individual components. Qualitative analysis of commercial products is frequently carried out by simple comparison of spectra, preferably after extraction of various additives, with the published spectrum of the suspected compound [2,26,27]. In polymers with more or less known chemical structures, infrared analysis can supply a great deal of other information about the physical state of the polymer, i.e., crystallinity, orientation, and microstructure [2,4,25,26].

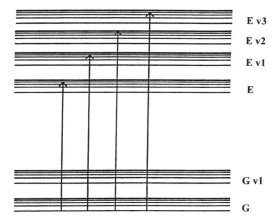

Fig. 14.4 Vibrational and rotational energy levels of a polyatomic molecule.

Quantitative Analysis

Band intensities are expressed as either transmittance (T) or absorbance (A). Transmittance is the ratio of the radiant power transmitted by a sample to the radiant power incident on the sample. Absorbance is the base 10 logarithm of the reciprocals of the transmittance: $A = \log_{10} T$. Quantitative measurements in the infrared usually begin with Beer's law and its analogs:

$$A = \varepsilon bc,$$

where ε is the molar absorptivity, b is the cell length, and c is the molar concentration of the absorbing substance. Figure 14.5 is the (a) transmittance and (b) absorbance spectrum of poly(vinyl chloride).

Problems are associated with quantitative analysis using IR. First, deviation from Beer's law affect quantitative analyses profoundly, especially those deviations resulting from saturation effects. Variations in the path length that are not accounted for can also cause problems. Second, specific interactions between components in the sample can influence the quantitation, especially those interactions that are temperature and pressure sensitive. Third, if the quantitation is based on the peak being due to only one absorbance when in reality it is a result of overlapping bands, then there will be a bias in data that is not necessarily linear. Currently available IR spectrometers have software packages containing matrix methods that simplify the operations associated with multicomponent

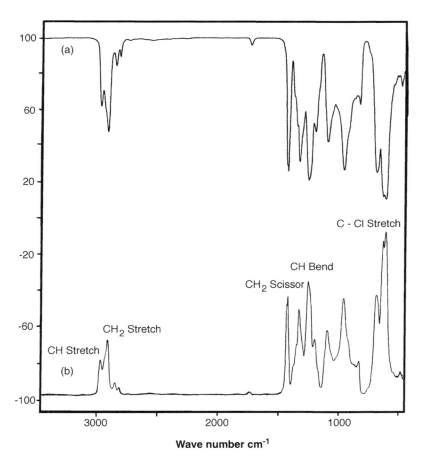

Fig. 14.5 Infrared spectrum of poly(vinyl chloride): (a) transmittance and (b) absorbance.

analysis [26]. If deviations of Beer's law occur, but the law of additivity still holds, correlation programs or statistical evaluation software packages such as least-squares regression, partial least-squares regression, and principle component analysis can be used during data processing and curve fitting [26].

Studies of the Chemical Nature of Polymers

Polymerization Kinetics and Cure Studies [2,4,25] Infrared spectra of monomers differ markedly from spectra of the polymers [2]. As a consequence, it is possible to use infrared spectroscopy to follow the course of polymerization reactions and to simultaneously analyze the structure of the polymer [2].

Any chemical reaction, whether involving the main chain or side groups, results in a change of composition of one or more groups and consequently in the IR spectrum. This makes it possible to study oxidation, thermal degradation, cyclization, grafting, and other reactions of polymers [2,4]. Evaluation of both qualitative and quantitative changes, as well as determination of kinetic constants of the reaction, is possible [2].

Average Molecular Weight [2] The average molecular weight can be determined from infrared spectra, provided that the polymer molecules have easily detectable end groups [2]. The lower the molecular weight, the more prominent are the absorption bands of the end groups. This measurement is of a quantitative nature, and the main problems are the determination of absorptivities, elimination of nonspecific absorptions, and interference of the absorption bands [2].

Studies of the Physical Nature of Polymers

Crystallinity In crystallization of polymers, the polymer forms "crystalline" and "amorphous" regions [2,4,25]. The formation of crystalline regions is accompanied by an increase in new vibrational modes caused by their crystal lattice interactions [2]. The IR spectrum of a given polymer differs by various absorption bands, depending on whether it is in the amorphous or crystalline state [2]. The IR spectrum exhibits "regularity" bands, splitting, and frequency shifts. Other absorption bands are not affected by crystallization and remain the same in both cases. Crystalline and amorphous bands can be used in the determination of the degree of crystallinity; independent bands are useful for the determination of sample thickness [2].

Molecular Orientation Characterization of molecular orientation is important as many physical and mechanical properties of polymers depend on the extent and uniformity of the orientation [2,4,25]. Orientation can be measured by using a variety of techniques [2,4,25,33,34]. IR spectroscopy not only allows the characterization of amorphous and crystalline phases separately, it also provides morphological data and can be used to map orientation with high spatial resolution [35].

Sample Preparation Techniques Applied to Polymers

There are three primary categories of IR techniques: transmission, reflectance, and photoacoustic. Transmission is the oldest and best known method for obtaining infrared spectra [23]. Gases, liquids, and solids can all be conveniently handled by transmission methods. The preferred method for polymers is to prepare thin films by either melt pressing or solvent casting. Care must be taken with either of these preparations as they can alter the crystallinity of a semicrystalline polymer or induce or destroy molecular orientation. Transmission techniques with a linear polarizer are also well suited to studying uniaxial orientation in polymer films. By changing the angle of incidence and polarizer orientation, information in the polymer chain axis direction in a drawn thin film can be obtained [36]. Brittle materials, such as cross-linked systems, can be ground to a fine powder and a KBr pellet [37] or Nujol Mull can be prepared [35].

Many polymers are too tough to be ground even at liquid nitrogen temperatures. Consequently, surface techniques are often used. Internal reflectance or attenuated total reflectance (ATR) is the second most commonly used infrared technique [38–40]. For soft or pliable polymers or solutions, ATR is an extremely versatile technique and the spectrum is similar to a transmission spectrum. Unlike transmission, the spectrum obtained is independent of sample thickness.

Infrared microscopy has become a useful technique for analyzing small samples such as surface defects, forensic samples, or trace contaminants in multilayer laminates [4,26]. The primary advantages of IR microscopy are not only related to improved optical and mechanical design, but also to manipulative capability [26].

APPLICABILITY

Infrared spectroscopy is a relatively simple technique, nondestructive, and versatile enough to analyze solids, liquids, and gases with a minimum of sample preparation. Polymers can be mixed with potassium bromide and then pressed into pellets. Films can be prepared from melt or cast from solution and can be studied easily. In bulk samples or powders, or if a concentration profile is needed, the reflectance technique is probably more suitable than transmission.

ACCURACY AND PRECISION

With most infrared spectrometers, it is possible to determine absorbance with a precision of 0.5 to 1.0%. However, the accuracy of the quantitative determination can vary from <1 to >10%, depending on the system analyzed [41].

SAFETY PRECAUTIONS

If a Fourier transform infrared spectrometer is being used, do not attempt to adjust the laser. Safety glasses must be worn in the laboratory at all times. Appropriate safety gloves and other personal protection equipment should be used when handling chemicals.

Material safety data sheets (MSDS) for all chemicals being used must be read prior to beginning the experiment. All chemicals should be considered hazardous from a standpoint of flammability and toxicity.

APPARATUS

1. Infrared spectrometer, with good resolution in the mid-IR region
2. IR salt plates
3. Balance capable of reading to a milligram
4. Aluminum weighing dishes
5. Stop watch
6. Desiccator
7. Spatula
8. Thermometers
9. Protective film strips for the IR salt plates

REAGENTS AND MATERIALS

1. Poly(Bisphenol A-co-epichlorohydrin)-glycidyl end capped, CAS [25036-25-3], Average molecular wt 348, m.p. 41–44°C, d 1.169.
2. Diethylenetriamine, 99%, CAS[111-40-0], molecular weight 103.17, m.p. −35°C, d 0.955.
3. Triethylenetetramine, 60%, CAS[112-24-3], molecular weight 146.24, m.p. 12°C, d 0.982.

PROCEDURE

The instrument manual and Ref. 42 to 45 should be read prior to the start of the experiment.

Note: Steps 1–16 and 17–19 should be run concurrently.

1. Place a small amount of the diethylenetriamine between the IR salt plates.
2. Record and plot the IR spectrum of diethylenetriamine.
3. Repeat steps 1 and 2 for the triethylenetetramine.
4. Weigh out 10.44 g of Bisphenol A epoxy resin into an aluminum weighing dish.
5. Using another aluminum weighing dish, weigh out 2.06 g of diethylenetriamine.
6. Add the diethylenetriamine to the aluminum weighing dish containing the Bisphenol A epoxy resin.
7. Record the time. This is time = 0.
8. Mix well with the spatula.
9. Quickly place a small amount of the mixture between two IR salt belts and place inside the IR spectrometer. (Note: To protect the salt plates use the protective film strips).
10. Acquire and plot the IR spectrum.
11. Record the time. This is time = t_1.
12. Place a thermometer into the remaining mixture.
13. Keep recording the temperature as you record the IR spectrum.
14. Repeat steps 10 and 11 every 15 min to follow the chemical reaction over the next 1.5 hr.
15. Store the samples in the desiccator until the next laboratory period.
16. Repeat steps 10 and 11.
17. Weigh out 6.49 g of Bisphenol A epoxy resin into an aluminum weighing dish.
18. In another aluminum weighing dish, weigh out 1.46 g of triethylenetetramine.
19. Repeat steps 6 to 14.

FUNDAMENTAL EQUATION

Beer's Law

$$A = \varepsilon bc$$

CALCULATIONS

1. Calculate the fundamental frequency expected in the infrared spectrum for the C—O stretching frequency. The value of the force constant is 5.0×10^5 dyne cm^{-1}.

2. From the infrared spectra acquired, record A at 3226 cm^{-1} and 912 cm^{-1}. Calculate ΔA, which is $A_0 - A_t$. Plot log A versus time, t, for each amine. Which amine is the most reactive?
3. Plot the recorded temperatures versus time for each reaction. Was there an exotherm? What was the maximum temperature reached during the course of the reaction?

REPORT

1. Describe the apparatus and experiment in your own words. Include a description of the infrared spectrometer and its mode of operation.
2. Tabulate total absorbance; baseline absorbance, and their difference (ΔA) for each monomer and the reaction product.
3. Describe the reactions taking place between the epoxy resin and the amines. Use IR spectra in your discussion.
4. Include spectra, tables, and plots.

ACKNOWLEDGMENTS

The authors thank Drs. Dana Garcia and Paul Chabot, Elf Atochem North America, for proofreading and for providing figures.

REFERENCES

1. I. N. Levine, "Molecular Spectroscopy," Chap. 6, Wiley, New York, 1975.
2. "Characterization of Polymers: Encyclopedia Reprints" (N. M. Bekalis, ed.), pp. 125–148. Wiley-Interscience, New York, 1971.
3. H. W. Thompson and P. Torkington, *Trans. Faraday Soc.* **41,** 246 (1945).
4. S. L. Hsu, in "Comprehensive Polymer Science: The Synthesis, Characterization, Reactions and Applications of Polymers" (G. Allen and J. C. Bevington, eds.), Vol. 1, Chap. 20, Pergamon Press, New York, 1989.
5. J. G. Grasselli, S. E. Mocadlo, and J. R. Mooney, in "Applied Polymer Analysis and Characterization: Recent Developments in Techniques, Instrumentation, Problem Solving" (J. Mitchell, Jr., ed.), Chap. III-A, Hanser Publishers, New York, 1987.
6. J. B. Bates, *Science* **191,** 31 (1976).
7. E. D. Becker and T. C. Farrar, *Science* **178,** 361 (1972).
8. J. A. De Haseth, "Fourier Transform Infrared Spectrometry," p 387 in "Fourier, Hadamard and Hilbert Transformation in Chemistry" (A. G. Marshall, ed.) Plenum, New York, 1982.
9. J. R. Ferraro and L. J. Basile, eds., "Fourier Transform Infrared Spectroscopy," Academic Press, New York, 1978.
10. J. G. Graselli and L. E. Wolfram, *Appl. Opt.* **17,** 1386 (1978).
11. J. G. Graselli, P. R. Griffiths and R. W. Hannah, *Appl. Spectrosc.* **36,** 81 (1982).
12. P. R. Griffiths, ed., "Transform Techniques in Chemistry," Plenum, New York, 1975.
13. P. R. Griffiths, "Chemical Infrared Fourier Transform Spectroscopy," Wiley, New York, 1978.
14. P. R. Griffiths, H. J. Sloane, and R. W. Hannah, *Appl. Spectrosc.* **31,** 485 (1977).
15. P. R. Griffiths, C. T. Foskett, and F. Curbelo, *Appl. Spectrosc. Rev.* **6,** 31 (1972).
16. P. R. Griffiths, *Appl. Spectrosc.* **31,** 497 (1977).
17. R. O. Kagel and S. T. King, *Ind. Res.* **15,** 54 (1973).
18. J. L. Koenig, *Am. Lab.* **6,** 9 (1974).
19. J. L. Koenig, *Appl. Spectrosc.* **29,** 293 (1975).
20. J. L. Koenig and D. L. Tabb, *Can. Res. Dev.,* September/December (1974).

21. A. L. Smith, "Applied Infrared Spectroscopy," Wiley, New York, 1979.
22. C. Y. Liang, in "Newer Methods of Polymer Characterization" (B. Li, ed.), Chap. 2, Interscience Publishers, 1964.
23. "Fourier Transform Infrared Characterization of Polymers" (H. Ishida, ed.), Plenum, New York, 1987.
24. R. M. Silverstein, G. C. Bassler, and T. C. Morrill, "Spectrometric Identification of Organic Compounds," 3rd Ed., Wiley, New York, 1974.
25. J. M. Chalmers and N. J. Everall, in "Polymer Characterization" (B. J. Hunt and M. I. James, eds.), Chap. 4, Balckie Academic and Professional, New York, 1993.
26. C.-P. Sherman, in "Handbook of Instrumental Techniques for Analytical Chemistry" (F. Settle, ed.), Chap. 15. Prentice Hall PTR, New Jersey.
27. Catalog of Infrared Spectrograms, Sadtler Research Laboratories, Philadelphia, PA, spectra indexed by name and by major bands.
28. G. Herzberg, "Infrared and Raman Spectra of Polyatomic Molecules," Van Nostrand, New York, 1945.
29. E. B. Wilson, J. C. Deceus, and P. C. Cross, "Molecular Vibrations," McGraw Hill, New York, 1955.
30. N. B. Colthup, L. H. Daly, and S. E. Weberly, "Introduction to Infrared and Raman Spectroscopy," Academic Press, Boston, 1990.
31. L. Woodward, "Introduction to the Theory of Molecular Vibrations and Vibrational Spectroscopy," Oxford University Press, Oxford, 1976.
32. F. A. Cotton, "Chemical Applications of Group Theory," 2nd Ed., Wiley-Interscience, New York, 1971.
33. R. J. Samuels, "Structured Polymer Properties," Wiley-Interscience, New York, 1974.
34. G. Wilkes, *Adv. Polym. Sci.* **8,** 91 (1971).
35. H. H. Willard, L. L. Merritt, Jr., and J. A. Dean, "Instrumental Methods of Analyses," 5th Ed., Van Nostrand, New York, 1974.
36. A. Cunningham, G. R. Davies, and I. M. Ward, *Polymer* **15,** 743 (1974).
37. O. Y. Ataman and H. B. Mack, *Appl. Spectrosc. Rev.* **13,** 1 (1977).
38. N. J. Harrick, "Internal Reflection Spectroscopy," Wiley, New York, 1967.
39. P. A. Wilkes, Jr., *Am. Lab.* **4,** 42 (1972).
40. B. Crawford, T. G. Goplen, and D. Swanson, in "Advances in Infrared and Raman Spectroscopy," Vol. 4, Chap. 2, Heyden, London, 1978.
41. E. A. Collins, J. Bareš, and F. W. Billmeyer, Jr., "Experiments in Polymer Science," Wiley, New York, 1973.
42. ASTM E 168-92, "Standard Practice for General Techniques of Infrared Quantitative Analysis," 1992.
43. ASTM E 1252-94, "Standard Practice for General Techniques for Qualitative Analysis," 1994.
44. ASTM E 1421-94, "Standard Practice for Describing and Measuring Performance of Fourier Transform Infrared (FT-IR) Spectrometers: Level Zero and Level One Tests," 1994.
45. ASTM E 334-90, "Standard Practice for General Techniques of Infrared Microanalysis," 1990.

EXPERIMENT 15

Thermogravimetric Analysis

INTRODUCTION

Thermogravimetric analysis (TGA) uses heat to drive reactions and physical changes in materials [1]. TGA provides a quantitative measurement of any mass change in the polymer or material associated with a transition or thermal degradation [2–17]. TGA can directly record the change in mass due to dehydration, decomposition, or oxidation of a polymer with time and temperature. Thermogravimetric curves are characteristic for a given polymer or compound because of the unique sequence of the physicochemical reaction that occurs over specific temperature ranges and heating rates and are a function of the molecular structure. The changes in mass are a result of the rupture and/or formation of various chemical and physical bonds at elevated temperatures that lead to the evolution of volatile products or the formation of heavier reaction products. From TGA curves, data concerning the thermodynamics and kinetics of the various chemical reactions [17–24], reaction mechanisms and the intermediate and final reaction products are obtained. Other processes that can be studied by TGA are adsorption and desorption phenomena, reactions with purge gases, ash content analysis, quantitative determination of additives (including plasticizers in polymers), solid-state reaction composition of filled polymers, rates of evaporation, and sublimation. TGA has also been used to estimate the flame retardancy of polymers, as enhanced flame retardancy is often paralleled by increased amounts of residual char at high temperature [25]. Similarly, antioxidant effectiveness can be gauged by the degree to which the degradation of the polymer is pushed to higher temperatures in the air purge. The temperature at which 5% of the starting mass has been lost is a convenient benchmark for comparing antioxidant efficiency.

PRINCIPLE

By definition [26,27], thermogravimetric analysis is a technique in which the mass of a substance is measured as a function of time or temperature while the substance is subjected to a controlled temperature program. Because mass is a fundamental attribute of a material, any mass change is more likely to be associated with a chemical change, which may, in turn, reflect a compositional change.

The sample is placed in a furnace while being suspended from one arm of a precision balance. The change in sample weight is recorded while the sample is either maintained isothermally at a temperature of interest or subjected to a programmed heating. The TGA curve may be plotted in either (a) the weight loss of the sample or (b) in differential form, (the change of sample weight with time) as a function of temperature (Fig. 15.1).

A TGA curve has limited ability to identify the sample or determine its composition if its nature is unknown. However, there are still many unique analytical applications based on TGA. Two primary applications are qualitative identification and compositional analysis.

Qualitative Identification

Different polymers have different thermal stabilities. An advantage of TGA is that it provides a rapid means to distinguish one polymer from another on the basis of the temperature range, extent, rate, and activation energy of decomposition. Figure 15.2 shows superimposed TGA curves of several polymers, where each has its own decomposition profile and can be distinguished from one another. A disadvantage of TGA is that many polymers do not show large enough decomposition differences to provide adequate resolution. Also, the thermal stability of a polymer can be affected by additives, previous heat treatment, and the inclusion of other substances. Given a TGA scan of an unknown sample, it is very difficult, if not impossible, to identify the material. However, TGA is very effective in a chemical laboratory or a processing plant where the sample system is well defined. One effective way to use TGA as a qualitative tool is to complement it with other analytical techniques.

Compositional Analysis

Additives

In known systems, TGA is useful for compositional analysis, especially for additives. TGA often proves effective in dealing with complex systems that are

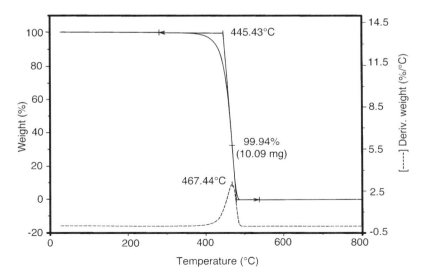

Fig. 15.1 Typical thermogravimetric mass loss of a sample of poly(ethylene), heated at a constant rate.

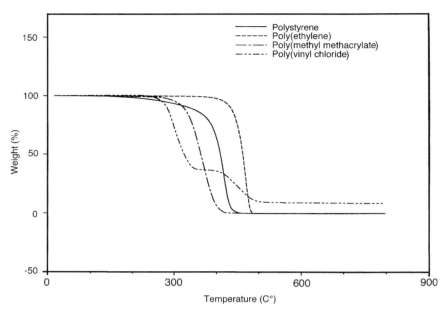

Fig. 15.2 Superimposed TGA curves of several polymers, ramp = 10°C/min, room temperature to 800°C, nitrogen gas purge.

difficult to analyze. TGA has been used in determining (a) plasticizers in poly(vinyl butyral) [28], (b) the moisture in nylon fibers [29], and (c) the glass content in an epoxy printed circuit board [29]. Thermal methods including TGA have been used in the rubber industry [30]. These complex formulations in either uncured master batches or cured vulcanizates consist mainly of oil, polymer, carbon black, and mineral filler. In most of these applications, the sample system is composed of volatile and nonvolatile components.

Polymer Thermal Stability: Degradation

There are three ways in which a polymer degrades: random chain scission, systematic chain scission, or a combination of the two. The degradation process often parallels the manner of polymer formation such that condensation polymers degrade more than chain growth (free radical) polymers. Random scission along the chain produces radicals or other reactive species. These reactive species may continue to break down into progressively smaller species, which become volatile and are lost, or they may attack other polymer chains leading to cross-linked polymers, which are less prone to degradation, and ultimately lead to a high temperature residue referred to as a char. A generalized degradation paralleling polymer formation is shown in Fig. 15.3.

A measure of the relative importance of these processes may be obtained by dividing the number of depolymerization steps occurring by the number of chain transfer and termination steps. This parameter is the zip length and gives a measure of the tendency of the polymer to break down into a monomer. If the zip length is low, there will be little monomer in the volatile products with an increased probability for a higher char yield.

Two factors favor high zip length: the absence of easily abstractable hydrogens and the resonance stabilization of the incipient radicals that can participate in monomer production. The following examples illustrate (see Fig. 15.4) the concept [5]. Both poly(methyl methacrylate) I and poly(methyl acrylate) II can generate radicals III and IV, where R represents part of the polymer chain,

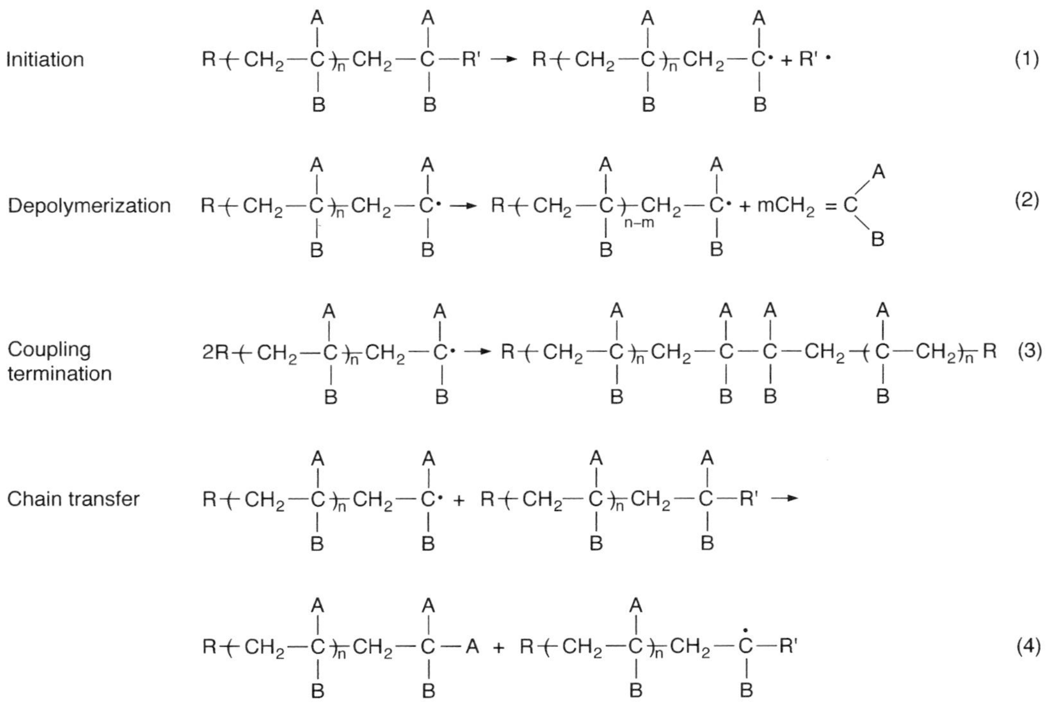

Initiation (1)

Depolymerization (2)

Coupling termination (3)

Chain transfer (4)

Fig. 15.3 Polymer degradation scheme.

which can be stabilized by resonance involving the carbonyl group. Poly(methyl acrylate) has an easily abstractable hydrogen (Fig. 15.4), unlike poly(methyl methacrylate) I. Consequently, poly(methyl methacrylate) has essentially 100 weight percentage yield of monomer compared to approximately 0.7 weight percentage for the acrylate.

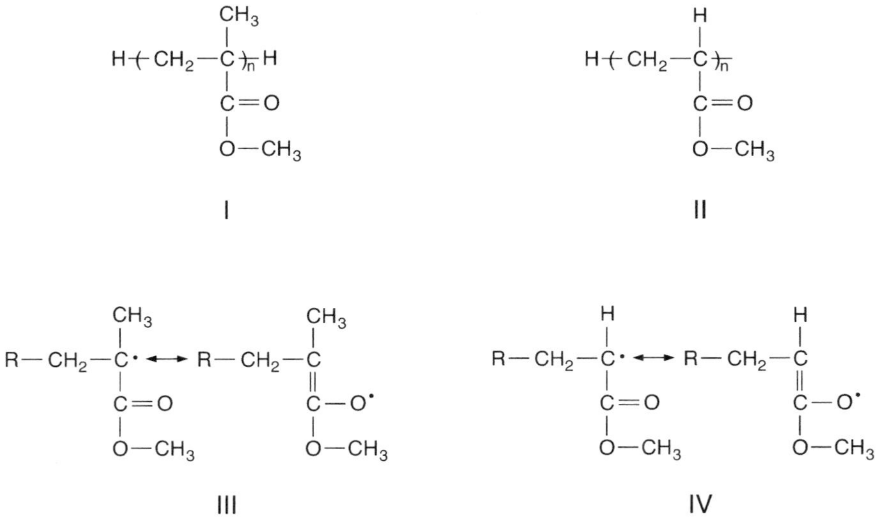

Fig. 15.4 Degradation schemes of poly(methyl methacrylate), I and III, and poly(methyl acrylate), II and IV.

Other degradation processes in addition to depolymerization can be initiated thermally. Thermal dehydrohalogenation of poly(vinyl chloride) is one such example [5].

$$R\text{-}(CH_2\text{-}\underset{Cl}{\overset{H}{C}})_n\text{-}CH_2\text{-}\underset{Cl}{\overset{H}{C}}\text{-}CH=CH_2 \longrightarrow R\text{-}(CH=\overset{H}{C})_n\text{-}CH=CH\text{-}CH=CH_2$$
$$+ (n+1)HCl$$

APPLICABILITY

The TGA technique can be applied to most materials that degrade due to instability brought on by increased temperature.

However, there are limitations to applying this technique to unknown materials. Because the recorded value is mass loss, this is not exactly a unique attribute. However, some interesting developments in combining TGA with other techniques, such as mass spectrometry, fourier transform infrared, or titrators [4,15], have expanded the utility of this technique to the identification of unknown materials.

ACCURACY AND PRECISION

Accuracy and precision vary with instrument model and total, initial, sample size, and sample preparation. A 0.05% accuracy and a precision of 0.1 μg can be achieved with a electrobalance [6]. Reproducibility between runs, except in relatively pure materials, is usually significantly poorer (see Notes section).

Absolute accuracy has no real meaning in this experiment; critical information consists of significant changes in mass loss or temperature of mass loss between samples. Inhomogeneity of samples, sample geometry, and sample size differences can have adverse effects on the reproducibility of data.

SAFETY PRECAUTIONS

Safety glasses should be worn in the laboratory at all times. Appropriate safety gloves should be used when handling chemicals. Material safety data sheets (MSDS) for all chemicals to be used must be read prior to beginning the experiment. All chemicals should be considered hazardous from the standpoint of flammability and toxicity.

The TGA apparatus becomes very hot and caution should be exercised while using it. In order to safely handle volatile materials or polymer decomposition products which may be irritating and harmful, it is suggested that a gas bubbler be constructed to scrub the gaseous effluent from the instrument (Fig. 15.5). Care must be taken to prevent excessive gas pressure from building up if the system is being operated under vacuum or with gases, both inert and flammable. Consult the instrument operations manual for any special safety instructions.

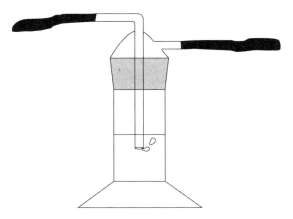

Fig. 15.5 A type of gas bubbler to absorb gases from the TGA experiment.

APPARATUS

1. Thermogravimetric analyzer unit, data acquisition system with a printer or plotter
2. Dewar flask
3. Gas bubbler or gas flow meter
4. Fifty (50 mg) calibrated weight standard or whatever is recommended by the instrument manufacturer
5. Appropriate regulator valves for gases
6. Effluent trap (Fig. 15.5) (see safety section)
7. Balance capable of weighing to 0.0001 g

REAGENTS AND MATERIALS

1. Sources of dry, high-purity nitrogen, air, and helium input to the TGA equipment
2. Twenty to thirty each of the following polymers:
   a. Poly(ethylene), high density (HDPE)
   b. Poly(ethylene), low density (LDPE)
   c. Poly(vinyl chloride) (PVC) 20–30 mg
   d. Poly(styrene) (PS)
   e. Poly(methyl methacrylate) (PMMA), syndiotactic
   f. PMMA, isotactic

PROCEDURE

It is recommended that Ref. 31 to 35 be read prior to beginning the experiments. Consult the operations manual for the instrument as to the proper means of:

a. Calibrating temperature and mass.
b. Loading and unloading sample (Notes 1 and 2).
c. Adjusting purge gas flow (Notes 3 and 4).
d. Operating program moble temperature controller (Note 5).
e. Setting-up data acquisition system.
f. Requiring data and data manipulation.

Experiment 15a. Polymer Fingerprinting

PURPOSE

The purpose of this experiment is to qualitatively identify an unknown polymer. Three polymer controls are run (PE, PS, and PMMA) which are chemically different. The unknown will have similar attributes. If the TGA apparatus has an autoloader, use of it will allow for unattended sequential running of pre-weighed samples.

1. Be sure the pans are clean and that the instrument has been tared for pan mass and calibrated using a standard mass in the range of the sample weight.
2. Weigh the pan. Record the weight in your notebook.
3. Add 10–20 mg of poly(styrene) to the pan and reweigh. Record the weights in your notebook.
4. Run the poly(styrene) sample on the instrument according to a linear programmed temperature ramp (usually 5, 10, or 20°C/min) to a final temperature greater than 600°C (see Fig. 15.6).
5. Plot data with the mass loss normalized to 100% on the ordinate axis and temperature on the abscissa.
6. Once the experiment is complete, cool the instrument to room temperature; remove, clean, and replace the sample holder; and retare the instrument in preparation for additional experiments. Using the same sample holder for the entire series of runs eliminates bouyancy problems.
7. Repeat steps 1 to 6 for the high-density poly(ethylene) and the poly(methyl methacrylate) sample.

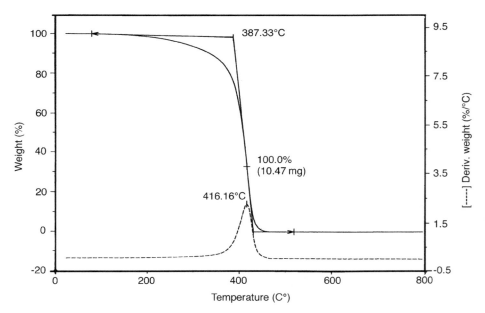

Fig. 15.6 TGA curve of poly(styrene) from room temperature to 800°C with a 10°C/min ramp under nitrogen gas purge.

8. Plot all data on one graph.
9. Repeat these procedures with the unknown polymer provided by the instructor.
10. Identify the polymer.

Experiment 15b. Polymer Degradation in Different Atmospheres

1. Be sure the pans are clean.
2. Weigh the pan. Record the weight in your notebook.
3. Add 10–20 mg of poly(vinyl chloride) to the pan and reweigh. Record the weight in your notebook.
4. Run the poly(vinyl chloride) in the instrument using an inert gas purge (usually N_2.)
5. Plot data with the mass loss normalized to 100% on the ordinate axis and temperature on the abscissa (see Fig. 15.7).
6. Once the decomposition of the sample is complete, cool the instrument to room temperature; remove, clean, and replace the sample holder; and retare the instrument in preparation for additional experiments. Using the same sample holder for the entire experiment eliminates bouyancy problems.
7. Repeat steps 1 to 5 using an oxygen-containing atmosphere (see Fig. 15.8).
8. Plot all data on one graph.

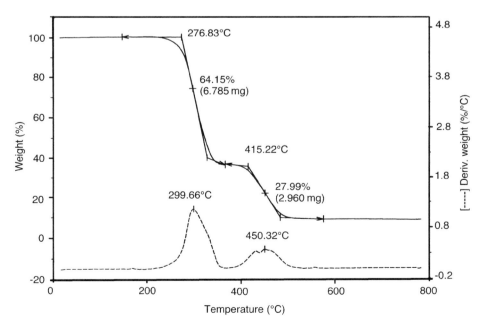

Fig. 15.7 TGA of poly(vinyl chloride) from room temperature to 800°C with a 10°C/min ramp under nitrogen gas purge.

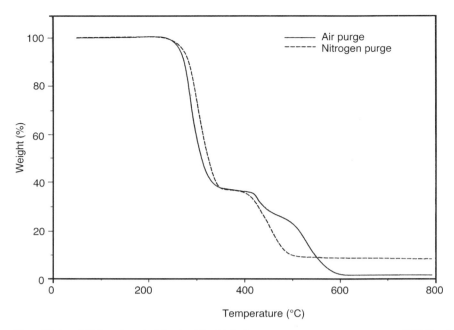

Fig. 15.8 TGA of poly(vinyl chloride) from room temperature to 800°C with a 10°C/min ramp using different purge gases.

FUNDAMENTAL EQUATIONS

$$\text{Percent mass loss} = 100(M_i - M_f)/M_i,$$

where M_i and M_f refer to the initial and final mass, respectively.

$$\frac{dm}{dt} = km^n,$$

which represents the rate of mass loss.

$$k = A \exp(-\Delta E/RT),$$

where A is the pre-exponential factor, E is the Arrhenius activation energy, R is the gas constant (8.134 J/mol K), and T is the temperature in degrees Kelvin [17,32].

CALCULATIONS

Calculations in TGA are based on the stoichiometry of the decomposition mechanism and may involve one or more reactions taking place simultaneously.

1. Calcium oxalate monohydrate ($CaC_2O_4 \cdot H_2O$) is a material widely used for the calibration of TGA equipment. It shows the following changes with increasing temperature (mass loss regions were obtained at a heating rate of 15°C/min):

$$CaC_2O_4 \cdot H_2O \xrightarrow{113-207°C} CaC_2O_4 + H_2O$$
$$CaC_2O_4 \xrightarrow{405-423°C} CaCO_3 + CO$$
$$3CaCO_3 \xrightarrow{626-793°C} 3CaO + 3CO_2$$

   a. What percent weight loss would be expected to accompany each of these changes?

b. Sketch the TGA thermogram from the above information for the transition of calcium oxalate monohydrate from 20 to 900°C.
2. A possible decomposition mechanism for poly(vinyl alcohol) is

$$-(CH_2-CH)_n\underset{OH}{|} \longrightarrow -(CH=CH)_n + nH_2O$$

a. If 10.65 mg of polymer is weighed out and there is a weight loss of 4.41 mg, is the reaction validated?
3. If 9.78 mg of poly(vinyl chloride) is weighed out and analyzed and a residue of 0.69 mg is found at 600°C, calculate the amount of char.

REPORT

For all experiments:
1. Describe the apparatus and experiment in your own words.
2. Report on the following for each material tested:
    a. The material name. If commercial, report the trade name, manufacturer, lot number, and composition.
    b. Purge gas and flow rate.
    c. Sample masses, temperature range studied and heating rate used.
3. Include the thermograms for all samples measured.
4. Report the temperature of mass loss onset and the temperature range of mass loss.
5. How do sample size and shape, gas pressure and atmosphere, and ignition of the sample affect the results?
6. What is the effect of the heating rate on the decomposition temperature and rate?
7. What effect can competitive reactions have on the thermogram?
8. How would you account for a weight gain in the thermogram?
9. How do sample size and sample form (geometry or packing) affect the profile of the decomposition?
10. Determine the Arrhenius activation energy, E, and the pre-exponential factor [17,32] for the depolymerization experiment.
11. What was the percent char in the PVC experiment for each of the purge gases? Explain any difference in the amount of char observed.

NOTES

Factors Affecting TGA Data

Several factors influence TGA data. Sample size and shape affect the rate and efficiency of decomposition. Powdered versus solid bulk samples will have different decomposition profiles due to the differing surface areas from which exiting decomposition products can leave the sample and be registered as mass losses. Similarly, the packing of the sample in the pan must be even and reproducible from run to run. Loosely distributed particles will heat more evenly and evolve volatilized products more evenly than mounded or densely packed samples. This can be especially important when looking at determinations of residual solvents, moisture or diffusion controlled losses such as plasticizer in the samples.

Quantity of sample is also an issue. The choice of sample mass should be such that signal-to-noise ratio is optimum. If the mass loss is very small, use of larger samples is helpful. However, if the sample size is too large, thermal gradients will occur which will retard the thermal decomposition profile. For many TGA systems, masses up to 50 mg are desirable.

The sampling atmosphere is an important factor. The sampling atmosphere may be an inert gas, or reactive gas or vacuum. When a gas is used, the atmosphere can either be flowing through the measurement chamber or static (i.e. nonflowing). If the atmosphere is static, the decomposition products may remain in the area of the sample and cause secondary reactions to occur, which are not necessarily related to the heating process.

The rate at which gas purge is swept across the specimen will affect results, due to the rate at which the exiting gas removes heat from the system (affecting control of heating ramps) and the rate at which volatilized products are removed.

The rate of heating is very important to obtaining good data. If the heating rate is too high, the sample temperature will not be linear with the increase of the furnace temperature. Thermal gradients in the sample will cause uneven decompositional behavior. Increasing the rate of heating will increase the temperature at which decomposition is seen; the curve will shift to the right, and differentiation between multiple mass loss regions may become ill-defined. Slower heating rates can be used to investigate overlapping regions of decomposition to separate loss regions. Isothermal conditions can be used to further investigate the effects of time and temperature on a reaction of interest.

ACKNOWLEDGMENTS

The authors thank Mrs. Kathy Lynn Lavanga, Rheometrics Scientific, Dr. Larry Judovits, Elf Atochem North America, Mr. Shih-Chien Chiu, Polytechnic University, and Dr. James S. Holton for proofreading and for providing corrections and figures.

REFERENCES

1. H. H. Willard, L. L. Merritt, Jr., and J. A. Dean, "Instrumental Methods of Analysis," 5th Ed., Van Nostrand, New York, 1974.
2. J. R. MacCallum, "Comprehensive Polymer Science: The Synthesis, Characterization, Reactions and Applications of Polymers" (G. Allen and J. C. Bevington, eds.), Vol. 1, Chap 37, Pergamon Press, New York, 1989.
3. B. Ki, "Newer Methods of Polymer Characterization" (B. Ki, ed.), Chap. IX, Interscience Publishers, New York, 1964.
4. J. Chiu, "Applied Polymer Analysis and Characterization: Recent Developments in Techniques, Instrumentation, Problem Solving" (J. Mitchell, Jr., ed.), Chap. 11-G, Hansen, New York, 1986.
5. E. M. Pearce, C. E. Wright, and B. K. Bordoloi, "Laboratory Experiments in Polymer Synthesis and Characterization," Pennsylvania State University, University Park, PA, 1982.
6. E. A. Collins, J. Bareš, and F. W. Billmeyer, Jr., "Experiments in Polymer Science," Experiment 24, Wiley-Interscience, New York, 1973.
7. P. D. Garn, "Thermoanalytical Methods of Investigation," Academic Press, New York, 1965.
8. W. W. Wendlandt, "Thermal Methods of Analysis," 2nd Ed., Wiley, New York, 1974.
9. T. Daniels, "Thermal Analysis," Kogan Page Ltd., London, 1973.
10. A. Blazek, "Thermal Analysis," Van Nostrand Reinhold, London, 1973.
11. R. Chen and Y. Kirsh, "Analysis of Thermally Stimulated Processes," Pergamon Press, London, 1981.
12. "Polymer Characterization by Thermal Methods of Analysis" (J. Chiu, ed.), Dekkar, New York, 1974.

13. "Thermal Characterization of Polymer Materials" (E. A. Turi, ed.), Academic Press, New York, 1981.
14. T. Hatakeyama and F. X. Quinn, "Thermal Analysis: Fundamentals and Applications to Polymer Science," Chap. 4, Wiley, New York, 1994.
15. R. E. Wetton, in "Polymer Characterization" (B. J. Hunt and M. I. James, eds.), Chap. 7, Blackie Academic and Professional, New York, 1993.
16. T. R. Crompton, "Analysis of Polymers: An Introduction," Chap. 6, Pergamon Press, New York, 1989.
17. M. E. Brown, "Introduction to Thermal Analysis: Techniques and Applications," Chapman and Hall, New York, 1988.
18. E. M. McCaffery, "Laboratory Preparation for Macromolecular Chemistry," Experiment 21, McGraw-Hill, New York, 1970.
19. R. M. Tuoss, I. O. Salyer, and H. S. Wilson, *J. Poly. Sci.* **A2,** 3147 (1964).
20. J. H. Flynn and L. A. Wall, *J. Res. N. B. S.* **70A,** 487 (1966).
21. F. L. Friedman, *J. Poly. Sci.* **B-7,** 41 (1969).
22. E. S. Freeman and B. Carroll, *J. Phys. Chem.* **62,** 394 (1958).
23. L. Reich and D. W. Levi, "Encyclopedia of Polymer Science and Technology" (H. F. Mark, N. E. Gaylord, and N. M. Bikales, eds.), Vol. 14, p. 1, Interscience, New York, 1971.
24. C. D. Doyle, "Techniques and Methods of Polymer Evaluations" (P. E. Slade, Jr. and L. T. Jenkins, eds.), Vol. 1, Chap. 4, Dekkar, New York, 1966.
25. D. W. Van Krevelen, *Polymer* **16,** 516, (1975).
26. R. C. Mackenzie, *Talanta* **16,** 1227, (1969).
27. R. C. Mackenzie, *J. Therm. Anal.* **13,** 13, (1978).
28. J. Chiu, *Appl. Polym. Symp.* **No. 2,** 25, (1966).
29. W. P. Brennan, *Thermochem. Acta* **18,** 101 (1977).
30. D. W. Brazier, *Rubber Chem. Technol.* **53,** 437 (1980).
31. ASTM E 1582-93.
32. ASTM E 1641-94.
33. ASTM E 698.
34. ASTM E 1131.

EXPERIMENT 16

Differential Scanning Calorimetry

INTRODUCTION

Thermal analysis is a term used to cover a group of techniques in which a physical property of a substance and/or its reaction product(s) is measured as a function of temperature. This experiment is confined to the area of differential thermal analysis (DTA) and, more specifically, its quantitative development, differential scanning calorimetry (DSC) [1–15].

DTA/DSC curves reflect changes in the energy of the system under investigation—changes that may be either physical or chemical in origin. DSC measures the heat required to maintain the same temperature in the sample versus an appropriate reference material in a furnace [9]. Enthalpy changes due to a change of state of the sample are determined. DTA differs from DSC in that the temperature difference is determined, rather than enthalpy differences between the sample and the reference material [9].

A number of important physical changes in a polymer may be measured by DSC. These include the glass transition temperature (T_g), the crystallization temperature (T_c), the melt temperature (T_m), and the degradation or decomposition temperature (T_D). Chemical changes due to polymerization reactions, degradation reactions, and other reactions affecting the sample can be determined (Table 16.1). A typical DSC trace showing these transitions is shown in Fig. 16.1.

BACKGROUND

There are two types of differential scanning calorimeters: (a) heat flux (ΔT) and (b) power compensation (ΔP). Subsequent sections of this experiment will not distinguish between the two types. In either type of calorimeter, the measurement is compared to that for a reference material having a known specific heat [16,17]. As ΔT and ΔP have opposite signs there is some potential for confusion [3], e.g., at the melting point, T_m, $T_s < T_r$, and $\Delta T < 0$, whereas $P_s > P_r$ and $\Delta P > 0$ because latent heat must be supplied (subscripts s and r refer to the sample and the reference material, respectively) [3].

Theory

This section considers only the simplest mathematics of DSC, where the temperature gradient within the sample can be neglected without serious error [18–21]. More complex mathematical treatment may be found elsewhere [21–23].

TABLE 16.1
Specific DTA and DSC Applications in Chemistry

Materials	Types of studies
Polymeric materials	Glass transition determinations
Organic materials	Decomposition reaction
Biological materials	Reaction kinetics
Carbohydrates	Phase diagrams
Amino acids and protons	Dehydration reactions
Pharmaceuticals	Solid-state reactions
Wood and related substances	Heats of absorption
Lubricating greases	Heats of reaction
Fats and oils	Heats of polymerization
Coal and lignite	Heats of sublimation
Natural products	Heats of transition
Catalysts	Catalysis
Coordination compounds	Radiation damage
Metal salt hydrates	Desolvation reactions
Metal and nonmetal oxides	Solid–gas reactions
Metals and alloys	Thermal stability
Clays and minerals	Oxidative stability
Soil	Curie point determinations
	Purity determinations
	Sample comparison

In DSC, the heat flow into the sample holder can be approximated by [3]:

$$dQ/dT = K(T_b - T),$$

where K is the thermal conductivity of the thermal-resistance layer around the sample, assumed to be dependent on geometry but independent of the temperature; T is the sample temperature; and T_b is the programmed block temperature:

$$T_b = T_0 + qt, \tag{1}$$

where T_0 is the initial temperature and q is the programmed heating rate. A similar equation may be written for the reference material.

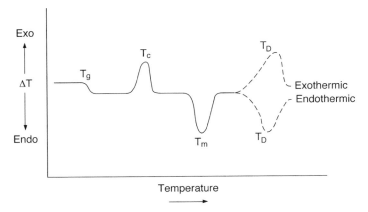

Fig. 16.1 A generic DSC curve depicting several transition types.

Heat Capacity

The basis for all calorimetric measurements is the determination of heat capacities. In the absence of any other transition, the DSC curve represents the change in the heat capacity of the sample over the experimental temperature range [5,24]. Detailed descriptions of experimental procedures and data treatment for using DSC to measure heat capacities are available [1,2,5,25–29] A simplistic approach is given below.

The heat absorbed on heating a sample with constant heat capacity, C_p, between temperatures T_0 and T is, by definition,

$$Q = C_p(T - T_0). \tag{2}$$

The basic equation for DSC can be derived from Eqs. (1) and (2) [30,31]

$$\Delta T = q\, C_p/K, \tag{3}$$

where ΔT is the difference in temperature between the reference material and the sample. The calibration factor K, thermal conductivity, is not usually calculated as a specific component of Eq. (4) and should be independent of temperatures for both power compensation (theoretical requirement) and heat flux (of successfully linearized) DSC [3,8]. The variation of K as a function of temperature and from day to day gives a good indication of instrument performance [3]. If the mass, m, of the sample is in grams, then

$$C_p = mc_p, \tag{4}$$

where c_p is the specific heat.

Enthalpy

The area of a DSC peak can be used to estimate the enthalpy of transition, ΔH, provided the thermal history of the sample is considered [29]. Calibration with respect to enthalpy requires an area that corresponds to a well-defined enthalpy change—a heat of fusion $\Delta H (T_m)$ is commonly used, especially that of indium [3].

Considering only the assumption of no difference in the heating rates for the sample and the reference and that the DSC curve then returns to the original baseline after the transition, then the enthalpy can be described in the following way:

$$\Delta H = \int_{T_i}^{T_f} C_p dT = \int_{T_i}^{T_f} (K\Delta T/q)\, dT, \tag{5}$$

where the T_i and T_f refer to the initial and final temperatures of the transition. The mathematical treatment of the more complicated can be found elsewhere [21–23].

Glass Transition

From a thermodynamic and mechanical point of view, the glass transition, T_g, is one of the most important parameters for characterizing a polymer system [1–3,24,29,32,33]. Consequently, the determination of the T_g is usually one of the first analyses performed on a polymer system.

A polymer may be amorphous, crystalline, or a combination of both. Many polymers actually have both crystalline and amorphous regions, i.e., a semicrystalline polymer. The T_g is a transition related to the motion in the amorphous regions of the polymer [3,8,9]. Below the T_g, an amorphous polymer can be said to have the characteristics of a glass, while it becomes more rubbery above the T_g [9]. On the molecular level, the T_g is the temperature of the onset of motion of short chain segments, which do not occur below the T_g [3,8,9,34–38].

The glass transition temperature can be measured in a variety of ways (DSC, dynamic mechanical analysis, thermal mechanical analysis), not all of which yield the same value [3,8,9,24,29]. This results from the kinetic, rather than thermodynamic, nature of the transition [40,41]. T_g depends on the heating rate of the experiment and the thermal history of the specimen [3,8,9]. Also, any molecular parameter affecting chain mobility effects the T_g [3,8]. Table 16.2 provides a summary of molecular parameters that influence the T_g. From the point of view of DSC measurements, an increase in heat capacity occurs at T_g due to the onset of these additional molecular motions, which shows up as an endothermic response with a shift in the baseline [9,24].

Melting and Crystallization

The most common applications of DSC are to the melting process which, in principle, contains information on both the quality (temperature) and the quantity (peak area) of crystallinity in a polymer [3]. The property changes at T_m are often far more dramatic than those at T_g, particularly if the polymer is highly crystalline. These changes are characteristic of a thermodynamic first-order transition and include a heat of fusion and discontinuous changes in heat capacity, volume or density, refractive index, birefringence, and transparency [3,8]. All of these may be used to determine T_m [8].

Generally, the crystalline melting point of a polymer corresponds to a change in state from a solid to a liquid and gives rise to an endothermic peak in the DSC curve [3,9]. The equilibrium melting point may be defined as

$$T_m = \frac{\Delta H_m}{\Delta S_m}, \tag{6}$$

where ΔH_m is the enthalpy of fusion and ΔS_m is the entropy of fusion [3,9]. In addition to determining the melting point and heat of fusion from DSC, the width of the melting range is indicative of the range of crystal size and perfection [3,9,24,29,52–54]. Because crystal perfection and crystal size are influenced by the rate of crystallization, T_m depends to some extent on the thermal history of the specimen [3,8,9].

Between the T_g and the T_m temperatures, another transition may be seen. The chains in crystallizable polymers have sufficient mobility so that ordering and crystallization may occur. The temperature at which this occurs is referred

TABLE 16.2
Structural Factors Affecting T_g [8]

Factors favoring a decrease in T_g	Factors favoring an increase in T_g
Main chain flexibility	Main chain rigidity
Flexible side chains	Bulky or rigid side chains [8,9]
Increased tacticity [42–44]	Increased cohesive energy density
Increased symmetry	Increased polarity
Addition of diluents or plasticizers [45]	Increased molecular weight [46–48]
Branching	Cross-linking [49–51]

to as the crystallization temperature, T_c, and represents a net output in energy resulting in an exothermic peak [9].

Above the T_m, the sample will decompose at its decomposition or degradation temperature, T_D. The decomposition process may be either exothermic or endothermic. A DSC curve of poly(vinyl chloride), containing different stabilizers, is shown in Fig. 16.2 and illustrates the various transitions.

Polymerization and Heats of Reaction

Any chemical reaction that is accompanied by enthalpy changes can be followed by calorimetric methods [3,24,29]. DSC is widely used to study polymerizing systems, especially epoxides [55–57]. Both the rate and the extent of reaction can be monitored using either isothermal or scanning modes of operation [3,24,29]. If volatile products are formed, the reaction must be carried out in sealed pans or under pressure [58] to conserve mass and avoid an uncertain correction for vaporization or to pressure-shift simultaneous vaporization events.

APPLICATIONS

Differential scanning calorimetry is applicable to the measurement of transition temperatures, specific heats, and heats of transition or reaction for all nonvolatile materials that do not evolve significant amounts of volatiles by reaction. The usual temperature range covered is −150 to 725°C.

ACCURACY AND PRECISION

It is essential that the sample be homogeneous and representative as milligram quantities of the sample are used.

Heats of transition and specific heats can be determined with a precision of about ±5%, with careful calibration and accurate weighing.

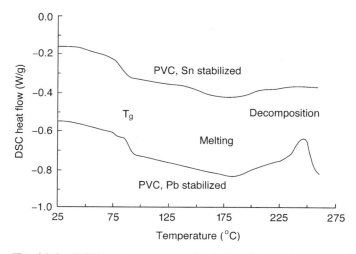

Fig. 16.2 DSC heat flow curve of poly(vinyl chloride), containing different stabilizers, heated at 20°C/min.

16. Differential Scanning Calorimetry

SAFETY PRECAUTIONS

Safety glasses must be worn in the laboratory at all times. A laboratory coat and appropriate gloves may be desirable. Material safety data sheets (MSDS), must be read for all chemicals used.

Normal safety precautions for laboratory work and for the use of electrical equipment, especially variable temperature accessories, must be observed. The thermal analysis experiment involves high temperatures and there is a danger of being burned. Consult instrument operating manual for specific cautions regarding operation.

If liquid nitrogen is used for experiments below room temperature, then the MSDS sheet or cryogenic liquid should be read. Avoid skin contact with liquid nitrogen or the supercooled gas, as a cryogenic "burn" will result. Do not put the liquid nitrogen in a sealed container that has no pressure relief device as the liquid nitrogen will become gaseous nitrogen and overpressurization of the container will occur; possibly resulting in an explosive depressurization.

APPARATUS

1. DSC equipment with a data acquisition system and a plotter or printer
2. DSC cell
3. Source of pure, low-pressure, dry nitrogen gas
4. Cooling accessory for below room temperature studies
5. Source of liquid nitrogen
6. Dewar flask
7. Pointed tip forceps
8. Balance capable of measuring to at least 0.00001 g
9. DSC aluminum sample pan and lids
10. Crimper

REAGENTS AND MATERIALS

1. Polymer samples, synthesized in earlier experiments (poly(styrene) from Experiment 3).
2. Commercial polymer samples. Suggested samples include poly(styrene), linear poly(ethylene), poly(ethylene terephthalate), poly(methyl methacrylate), or poly(vinyl chloride).
3. Glass beads or aluminum oxide, Al_2O_3 fine powder.

PREPARATION

1. One hour in advance of data acquisition, the equipment should be turned on.
2. The polymer samples should be free from impurities; including monomers, water, and solvent. The polymer sample should be homogeneous and of a large surface area (see Notes 1 and 2).

PROCEDURE

Control operations manual for instrument for proper methods of:

a. Temperature and heat calibration verification.
b. Sample pan type and crimping of the sample pan.

c. Loading sample in pan.
d. Operating temperature controller and setting temperature limits.
e. Setting-up and using data acquisition system.
f. Using the data analysis system.

Refs. 32 and 59 should be read before beginning the experiment.

First-Order Transition of Poly(amide)

Calibration

The DSC should be calibrated before analyzing the polymer. Calibration should be verified by using either indium or tin (Table 16.3). Temperature and heat values should be within normal ranges for the instrument [16,17] or at least within the precision and accuracy limits specified within Refs. 32 and 59.

Sample Preparation

1. Use forceps to handle the sample pan and lid. Obtain the tare weight of a sample pan and lid (the type that can be hermetically sealed is not required). (*Note:* Select pans with flat undistorted bottoms so that good contact with the cell platforms is assured.)
2. Record the mass of the sample pan and lid to five significant figures.
3. Add about 10 mg of polyamide or calibration standard to the sample pan and lid and weigh again. A powdered sample provides better thermal contact.
4. Record the mass of the sample to five significant figures.
5. Remove the polymer sample pan and lid carefully from the balance using forceps. (*Note:* Skin moisture and oils will be left on the sample pan and lid if they are picked up by fingers. This extra mass will effect the DSC experiment.)
6. Crimp the samples pan and lid closed.
7. Tare another sample pan and lid. This is the reference pan.
8. Record the weight of the sample to five significant figures.
9. Weigh out either 10 mg of glass beads or Al_2O_3 into the reference pan or just use an empty reference pan [9].
10. Crimp the reference pan and lid closed.
11. Purge the cell with nitrogen at a 30-ml/min gas flow rate.

Sample Analysis

12. Record the first thermal cycle (used for calibration verification) by heating the sample at a rate of 10°C/min under nitrogen atmosphere from ambient to 30°C above the expected melting point or glass transi-

TABLE 16.3
Typical DSC Standard Reference Materials

Standard	Melting point (°C)	Heat of fusion (kJ/kg)
Benzoic acid	122.4	142.04
Indium	156.4	28.45
Tin	231.9	59.50
Lead[a]	327.4	22.94
Zinc	419.5	102.24

[a] If lead is used as a standard, a fresh sample should be used each time.

tion point or up to a temperature high enough to erase previous thermal history (see Note 1).
13. Hold the temperature for 10 min.
14. Cool to 50°C below the peak crystallization temperature at a rate of 10°C/min.
15. Repeat heating as soon as possible under nitrogen at a rate of 10°C/min and record the heating curve.
16. Hold the temperature for 10 min.
17. Repeat cooling under nitrogen at a rate of 10°C/min and record the cooling curve.
18. Measure the temperatures for the following: T_f, T_m, T_c, and T_e, where T_f is the extrapolated onset temperature, T_m is the melting peak temperature, T_c is the crystallization peak temperature, and T_e is the extrapolated end temperature. Report two T_m values if observed.

FUNDAMENTAL EQUATIONS

Equilibrium Melting Point

$$T_m = \frac{\Delta H_m}{\Delta S_m}$$

Enthalpy

$$\Delta H/m = (K/mq) \int_{T_i}^{T_f} \Delta T \, dT$$

CALCULATIONS

1. Read K from a calibration curve obtained with materials of known heats of fusion.
2. Tabulate the extrapolated onset and peak or inflection temperatures for all transitions. (see Fig. 16.3).
3. A new polymer showed an endothermic transition at 80°C, a T_g of 145°C, a T_m of 235°C, and an exothermic decomposition starting at 400°C. Temperatures are given at the peaks. Sketch the thermogram.

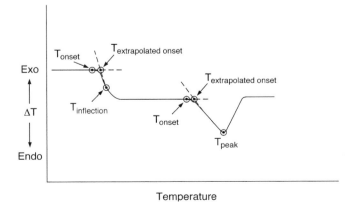

Fig. 16.3 Different ways of recording transition temperatures.

REPORT

1. Describe the apparatus and experiment in your own words.
2. Complete identification and description of the material tested, including its physical appearance. If it is a commercial sample, include its tradename, manufacturer, and lot number.
3. Describe the temperature calibration procedure.
4. Identify the sample atmosphere by pressure, gas flow, etc.
5. Record the results of the transition measurements using the temperature parameters (T_m, etc.).
6. Any side reaction should be reported and identified if possible; side reactions include cross-linking, oxidation, and thermal degradation.
7. What parameters effect the baseline and why?
8. Include all graphs from the instruments, for all experiments.

NOTES

1. The initial heating to a temperature about 30°C above the T_m or suggested upper temperature limit (cited in Refs. 32 and 59) and then cooled at a controlled rate. This annealing permits residual volatiles to leave as well as relieving mechanical stress and removes residual crystallinity.
2. The selection of temperature is critical. The time of exposure to high temperature should be minimized to avoid sublimation or decomposition. In some cases, the preliminary thermal cycle may interfere with the transition of interest, causing an incorrect transition or eliminating a transition.

ACKNOWLEDGMENTS

The authors thank Dr. Larry Judovits, Elf Atochem North America, Kathy Lynn Lavanga, Rheometric Scientific, and Dr. James S. Holton for proofreading, making corrections, and providing figures.

REFERENCES

1. "Thermal Characterization of Polymeric Materials" (E. A. Turi, ed.), 1st Ed., Academic Press, New York, 1981.
2. "Thermal Characterization of Polymeric Materials" (E. A. Turi, ed.) 2nd Ed., Academic Press, New York, 1997.
3. M. J. Richardson, "Comprehensive Polymer Science: The Synthesis, Characterization, Reactions and Application of Polymers" (G. Allen and J. C. Bevington, eds.), Vol. 1, Chap. 38, Pergamon Press, New York, 1989.
4. M. E. Brown, "Introduction to Thermal Analysis: Techniques and Applications," Chapman and Hall, New York, 1989.
5. B. Wunderlich, "Thermal Analysis," Academic Press, New York, 1990.
6. "Calorimetry and Thermal Analysis of Polymers" (V. B. F. Mathot, ed.), Hanser Publishers, New York, 1994.
7. B. Ke, "Newer Methods of Polymer Characterization" (B. Ke, ed.), Chap. IX, Interscience Publishers, New York, 1964.
8. E. A. Collins, J. Bareš, and F. W. Billmeyer, Jr., "Experiments in Polymer Science," Wiley, New York, 1973.

9. E. M. Pearce, C. E. Wright, and B. K. Bordoloi, "Laboratory Experiments in Polymer Synthesis and Characterization," The Pennsylvania State University, University Park, PA, 1982.
10. H. H. Willard, L. L. Merritt, Jr., and J. A. Dean, "Instrumental Methods of Analysis," 5th Ed., Van Nostrand, New York, 1974.
11. S. L. Boersma, *J. Am. Ceram. Soc.* **38,** 281 (1955).
12. M. J. O'Neill, *Anal. Chem.* **36,** 1238 (1964).
13. A. P. Gray, in "Analytical Calorimetry" (R. S. Porter and J. F. Johnson, eds.), Vol. 1, p. 209, New York, 1968.
14. R. A. Baxter, in "Thermal Analysis," Vol. 1, p. 65, Academic Press, New York, 1969.
15. S. C. Mraw, *Rev. Sci. Instrum.* **53,** 228 (1982).
16. ASTM E 967-92, "Standard Practice for Temperature Calibration of Differential Scanning Calorimeters and Differential Thermal Analyzers," 1992.
17. ASTM E 968-83, "Standard Practice for Heat Flow Calibration of Differential Scanning Calorimeters," 1983.
18. F. H. Müller and H. Martin, *Kolloid Z.* **172,** 97 (1960).
19. G. Adam and F. H. Müller, *Kolloid Z.-Z. Polym.* **192,** 29 (1963).
20. H. Martin and F. H. Müller, *Kolloid Z.-Z. Polym.* **192,** 19 (1963).
21. B. Wunderlich, "Techniques of Chemistry: Physical Methods of Chemistry" (A. Weissberger and B. W. Rossiter, eds.), Vol. 1, Chap. VIII, Wiley, New York, 1971.
22. P. D. Garn, "Thermoanalytical Methods of Investigation," Academic Press, New York, 1965.
23. W. J. Smothers and Y. Chiang, "Handbook of Differential Thermal Analysis," Chemical Publishing Co., New York, 1966.
24. A. J. Pasztor, in "Handbook of Fundamental Techniques for Analytical Chemistry" (F. Settle, eds.), Chap. 50, Prentice Hall PTR, New Jersey, 1997.
25. U. Gaur, A. Mehta, and B. Wunderlich, *J. Therm. Anal.* **13,** 71 (1978).
26. M. J. Richardson, in "Developments in Polymer Characterization" (J. V. Dawkins, eds.), Vol. 1, p. 205, Applied Science, London, 1978.
27. M. J. Richardson, in "Compendium of Thermophysical Property Measurement Methods" (K. D. Maglic, A. Cezairliyan, and V. E. Peletsky, eds.), Vol. 1, p. 669. Plenum Press, New York, 1984.
28. A. Xenopoulos and B. Wunderlich, *J. Polym. Sci. B Polym. Phys.* **28,** 2271 (1990).
29. T. Hatakeyama and F. X. Quinn, "Thermal Analysis: Fundamentals and Applications to Polymer Science," Wiley, New York, 1994.
30. A. P. Gray, in "Analytical Calorimetry" (R. S. Porter and J. F. Johnson, eds.), p. 209, Plenum Press, New York, 1968.
31. R. A. Baxter, in "Thermal Analysis: Instrumentation, Organic Materials and Polymers" (R. F. Schwenker and P. D. Garn, eds.), Vol. 1, p. 65, Academic Press, New York, 1969.
32. ASTM D 3418-82, "Standard Test Method for Transition Temperatures of Polymers by Thermal Analysis," 1982.
33. ASTM E 1356-91, "Standard Test Method for Glass Transition Temperature by Differential Scanning Calorimetry or Differential Thermal Analysis."
34. E. M. McCaffery, "Laboratory Preparation for Macromolecular Chemistry," Exp. 22, McGraw-Hill, New York, 1970.
35. F. W. Billmeyer, Jr., "Textbook of Polymer Science," 2nd Ed., Interscience, New York, 1971.
36. E. M. Barrall II and J. F. Johnson, in "Techniques and Methods of Polymer Evaluation" (P. E. Slade, Jr. and L. T. Jenkins, eds.), Dekker, New York, 1966.
37. D. J. David, in "Techniques and Methods of Polymer Evaluation" (P. E. Slade, Jr. and L. T. Jenkins, eds.), Dekker, New York, 1966.
38. D. Braun, H. Cherdon and W. Kern, "Techniques of Polymer Synthesis and Characterization," p. 24, p. 79, Interscience, New York, 1971.
39. B. Cassel and B. Twombly, "Materials Characterization by Thermomechanical Analysis" (A. T. Rega and C. M. Neag, eds.), ASTM STP 1136 (Philadelphia: American Society for Testing and Materials, 1991), 108–119.
40. A. R. Ramos, J. M. Hutchinson, and A. J. Kovacs, *J. Polym. Sci. Polym. Phys. Ed.* **22,** 1655 (1984).

41. C. T. Moynihan, A. J. Easteel, J. Wilder, and J. Tucker, *J. Phys. Chem.* **78,** 2673 (1974).
42. N. W. Johnston, *J. Macromol. Sci. Rev. Makromol. Chem.* **14,** 215 (1976).
43. P. R. Couchman, *Macromolecules* **15,** 770 (1982).
44. J. S. Roman, E. L. Madriega, and J. Guzman, *Polym. Commun.* **25,** 373 (1984).
45. L. H. Dunlap, C. R. Foltz, and A. G. Mitchell, *J. Polym. Sci. Polym. Phys. Ed.* **10,** 2223 (1972).
46. T. G. Fox and P. J. Flory, *J. Appl. Phys.,* **21,** 581 (1950).
47. R. F. Boyer, *Macromolecules* **7,** 142, (1974).
48. K. Veberreiter and G. Kanig, *J. Colloid. Sci.* **7,** 569 (1952).
49. F. Rietsch, D. Daveloose, and D. Froelich, *Polymer* **17,** 859 (1976).
50. J. H. Glans and D. T. Turner, *Polymer* **22,** 1540 (1981).
51. L. H. Judovits, R. C. Bopp, U. Gaur, and B. Wunderlich, *J. Polym. Sci. Polym. Phys. Ed.* **24,** 2725, (1986).
52. H. N. Beck, *J. Appl. Polym. Sci.* **19,** 371 (1975).
53. R. Legras, C. Bailly, M. Daumerie, J. M. Dekoninck, J. P. Mercier, V. Zichy, and E. Nield, *Polymer* **25,** 835 (1984).
54. D. Garcia, *J. Polym. Sci. Polym. Phys. Ed.* **22,** 2063 (1984).
55. R. B. Prime, in "Thermal Characterizations of Polymeric Materials" (E. A. Turi, ed.), p. 435. Academic Press, New York, 1981.
56. J. M. Barton, *Adv. Polym. Sci.* **72,** 111 (1985).
57. D. W. Brazier, in "Developments in Polymer Degradation" (N. Grassie, ed.), Applied Science, Vol. 3, p. 27, London, 1981.
58. P. F. Levy, G. Nieuweboer, and L. C. Semanski, *Thermochim. Acta* **1,** 429 (1970).
59. ASTM D 3417-83, "Standard Test Method for Heats of Fusion and Crystallization of Polymers by Thermal Analysis," 1983.

CHAPTER 17

Dilute Solution Viscosity of Polymers

INTRODUCTION

One of the most characteristic features of a dilute polymer solution is that its viscosity is considerably higher than that of the pure solvent. The large differences in size between solute and solvent molecules give rise to this effect. The change in viscosity can be significant even at low polymer concentrations, especially for polyelectrolytes and polymers with high molecular weights. Dilute solution viscometry is concerned with accurate quantitative measurement of the increase in viscosity and allows determination of the intrinsic ability of a polymer to increase the viscosity of a particular solvent at a given temperature [1–13]. This quantity provides information relating to the size of the polymers in solution, including the effects on chain dimensions of polymer structure, molecular shape, degree of polymerization, and polymer–solvent interactions. Most commonly, dilute solution viscosity is used to estimate the molecular weight of a polymer, which involves the use of semiempirical equations that have to be established for each polymer–solvent–temperature system by analysis of polymer samples whose molecular weights are known. The advantage of dilute solution viscosity is that it is simple, fast, and inexpensive. It is applicable over the complete range of attainable molecular weights. The disadvantage is that it provides estimates of molecular weight that are not absolute.

Solution viscosity is an excellent method for quality control for relatively uniform polymer samples. There is an ASTM test method for determining inherent viscosity (ASTM D 4603) that uses poly(ethylene terephthalate) and one for determining intrinsic viscosity of cellulose (ASTM D 1795) that describes a one-point method for estimating intrinsic viscosity. The result is useful as it relates viscosity to molecular weight, which is useful for checking different batches of polymer in a production line to help ensure uniformity.

PRINCIPLE

The frictional resistance of liquids to shear is characterized by the coefficient of viscosity, η, as defined by

$$\eta = \tau/\dot{\gamma}, \qquad (1)$$

where τ is the shearing stress per unit of surface and $\dot{\gamma}$ is the viscosity gradient perpendicular to the shearing stress. With τ in dynes/cm^2 and $\dot{\gamma}$ in sec^{-1}, η is given in poise. To determine accurately small viscosity increases brought about by dissolving small amounts of polymer in solvents, it is critical to measure viscosities at closely controlled temperatures.

The simplest experimental method for the determination of viscosity is the measurement of the time, t, required for the passage of a volume, V, through a capillary of length, ℓ, with a circular cross section of radius, r. The relationship between η and t is given by

$$\eta/\rho = At + B/t, \tag{2}$$

where ρ is the density of the liquid whereas A and B depend on the dimensions of the capillary. A is defined as

$$hg\,\pi\,r^4/8V\ell, \tag{3}$$

where h is the mean hydrostatic head of the fluid and g is the gravitational constant. The B term, the so-called "kinetic energy correction," arises from the back pressure produced by the deceleration of the fluid as it emerges from the capillary.

In interpreting the comparison of the viscosity of the pure solvent, η_ρ, and the viscosity, η, of a polymer solution, the terms in Table 17.1 are commonly used. In the equations in Table 17.1, the concentration, c, of the polymer is given in g/100 ml; however, IUPAC proposes c in g/ml. $[\eta]$ has the dimension c^{-1}, thus $[\eta]$ will be given in either dl/g or ml/g, depending on the units used for c. In a dilute solution, η_{sp}/c is linear in c and $[\eta]$ may be obtained by extrapolating a plot of η_{sp}/c versus c to $c = 0$.

The linearity of the plot of η_{sp}/c versus c and the dependence of the slope of the linear plot on $[\eta]$, which is the intercept, has been described by a purely empirical equation by Huggins.

$$\eta_{sp}/c = [\eta] + k'[\eta]^2 c, \tag{4}$$

where k', the Huggins constant, has values around 0.35 in strongly solvating solvents but may increase in poorly solvating media to values four times as large.

Another commonly used empirical equation is

$$(\ln\,\eta_r)/c = [\eta] + k''[\eta]^2 c, \tag{5}$$

where k'' is the Kraemer constant, which is usually negative. The relation between the Huggins constant and the Kraemer constant can be obtained as shown in the following equations. By definition,

$$\ln\,\eta_r = \ln\,(1 + \eta_{sp}) \tag{6}$$

$$\ln\,\eta_r = \eta_{sp} - \tfrac{1}{2}\,\eta_{sp}^2 + \ldots \tag{7}$$

TABLE 17.1
Viscometric Terms

Common name	IUPAC name	Symbol and defining equation
Relative viscosity	Viscosity ratio	$\eta_r = \eta/\eta_0 \simeq t/t_0$
Specific viscosity	—	$\eta_{sp} = \eta_r - 1 = (\eta - \eta_0)/\eta_0 \simeq (t - t_0)/t_0$
Reduced viscosity	Viscosity number	$\eta_{red} = \eta_{sp}/c$
Inherent viscosity	Logarithmic viscosity	$\eta_{inh} = (\ln\,\eta_r)/c$
Intrinsic viscosity	Limiting viscosity number	$[\eta] = (\eta_{sp}/c)_{c \to 0} = [(\ln\,\eta_r)/c]_{c \to 0}$

Upon substitution of η_{sp} in Eq. (7) into Eq. (4),

$$\ln \eta_r = [\eta]c + (k' - 1/2)[\eta]^2 c^2 \ldots \quad (8)$$

Comparing Eq. (8) with Eq. (5), neglecting terms having powers higher than 2 in c gives

$$k'' \simeq k' - 1/2. \quad (9)$$

Flexible Chains

Free-draining models were among the first to be considered [14–18]. For flexible polymer chains of sufficient length, $[\eta]$ behaves as if the polymer coil occupied a spherical volume through which the solvent cannot flow. Under these conditions,

$$[\eta] = \Phi \, (\bar{r}^2)^{3/2}/M, \quad (10)$$

where $(\bar{r}^2)^{3/2}$ is the root-mean-square separation of the two chain ends, M is the molecular weight of the polymer, and Φ is a universal constant estimated by Flory as 2.1×10^{21} if $(\bar{r}^2)^{3/2}$ is given in angstroms.

Evaluation of Molecular Weight

Flexible polymer chains expand with increasing solvent power of the medium, leading to an increase in $[\eta]$ with increasing polymer solvation. For chains of a similar kind, varying in length (homologous series), the relationship between $[\eta]$ and molecular weight, M, may be represented by the Mark–Houwink relationship [19–22].

$$[\eta] = KM^a, \quad (11)$$

where K and a are constants for a given polymer–solvent–temperature system. The exponent a increases with the solvent power of the medium. Theory predicts that it should lie in the range $0.5 < a < 0.8$ for flexible chains, $0.8 < a < 1.0$ for inherently stiff molecules (e.g., cellulose derivatives, DNA), and $1.0 < a < 1.7$ for highly extended chains (e.g., polyelectrolytes in solutions of very low ionic strength [1]). The value of K tends to decrease as a increases and for flexible chains it is typically in the range of 10^{-3} to 10^{-1} cm^3 g^{-1}. Generally, a plot of log $[\eta]$ against log [M] is fitted to a straight line from which log K and a are the intercept and slope, respectively. Other procedures for the evaluation of K and a have been proposed [23–28]. Comprehensive lists of K and a values are available [29–31] so that calibration is unnecessary for many polymers.

If Equation (11) represents the relationship between $[\eta]$ and M for a monodisperse polymer sample, then the intrinsic viscosity for a polydisperse sample containing weight fraction w_i with a molecular weight M_i will be

$$[\eta] = K \sum_{i=0}^{i} w_i M_i^a = K M_v^a \quad (12)$$

where M_v is referred to as the "viscosity average molecular weight."

As specific viscosities are additive in the limit of infinite dilution, the weight average intrinsic viscosity is obtained for a polydisperse polymer and $\overline{M}_v$ is defined as

$$\overline{M}_v = \left[\frac{\sum w_i M_i^a}{\sum w_i}\right], \quad (13)$$

where w_i is the total weight of molecules of molecular weight M_i. Consequently, $\overline{M}_v$ is closer to the weight average $\overline{M}_w$ than to the number average $\overline{M}_n$ molecular weight, and is equal to $\overline{M}_w$ when $a = 1$.

Branched Polymers

The effect of branching is to increase the segment density within the molecular coil. Thus a branched molecule occupies a smaller volume and has a lower intrinsic viscosity than a similar linear molecule of the same molecular weight. The degree of branching is often characterized in terms of the branching factor [1] in Eq. (14), where the subscripts B and L, respectively, refer to branched and linear polymers of the same molecular weight:

$$g' = \frac{[\eta]_B}{[\eta]_L}. \tag{14}$$

In order to estimate g', it is necessary to measure $[\eta]_B$ and $\overline{M}_{w,B}$ (or $\overline{M}_{n,B}$). Also, a Mark–Houwink relationship of the linear polymer must be known for the conditions employed in the measurement of $[\eta]_B$. In general, if the log $[\eta]$ versus log M relation is known for the linear species, the deviation from this relationship may be used to estimate the number of branches, provided certain assumptions can be made concerning the distribution of branches.

Copolymers

The presence of a second type of repeat unit causes the dilute solution behavior to be more complex than that of homopolymers [1]. Copyolymer composition and sequence distribution directly effect the intrinsic viscosity. Interactions between unlike chain segments and preferential interaction of solvent molecules with one of the comonomers are also of considerable importance.

Polyelectrolytes

Polyelectrolytes are most commonly studied as solutions in aqueous media as a consequence of their poor solubility in organic solvents. In order to minimize the effects of ionization in an aqueous media, evaluation of $[\eta]$ is done in an aqueous solution of an inert 1:1 electrolyte as the solvent. In such solvents, polyelectrolytes behave as if they were neutral polymers [32–34].

APPLICABILITY

The dilute solution viscosity measurement is applicable to all polymers that dissolve to give stable solutions at temperatures close to the boiling point of the solvent.

ACCURACY AND PRECISION

For most polymer–solvent systems, the reduced, inherent, and intrinsic viscosities can be determined to well within $\pm$ 0.01 dl/g.

SAFETY PRECAUTIONS

Safety glasses must be worn in the laboratory at all times. Appropriate safety gloves must be worn when preparing solutions. Caution must be used in making polymer solutions. Material safety data sheets must be read regarding the solvent, polymer, and suspected additives as many solvents and monomers are toxic, carcinogenic, or flammable. Polymer solutions should be made up at concentrations less than one weight percent in small quantities (<50 ml) in a hood or well-ventilated area. Also, many polymer solutions may have to be heated on a hot plate or sand bath in order to facilitate the polymer dissolving. Avoid the use of flames or sources of electrical sparks.

APPARATUS

1. Balance capable of reading to 0.001 g or preferably 0.0001 g
2. Constant temperature bath capable of maintaining ±0.01°C at 25°C
3. Heating unit for the bath
4. Ubbelohde viscometer with efflux times greater than 100 sec for the solvent
5. Timer, graduated in 0.1 sec or less, or stop watch
6. Sintered glass filter of fine porosity or, alternatively, 0.1-μm Millipore filter in a suitable holder
7. Five- and 10-ml volumetric pipettes
8. 100-ml volumetric flask, stoppered
9. Rubber suction bulb
10. Source of dry, filtered nitrogen gas
11. Weighing paper or aluminum dishes
12. Magnetic stirring and heating unit
13. 100-ml Erlenmeyer flask, stoppered
14. Sample holder and clamp to suspend the viscometer in the constant temperature bath

REAGENTS AND MATERIALS

1. Poly(styrene)
2. Toluene, reagent grade
3. Aqua regia cleaning solution (1 volume of conc. nitric acid to 3 volumes of conc. hydrochloric acid mixed in a hood)

PREPARATION

Prepare ahead of time the polymer solutions: poly(styrene) at 1 g in 100 ml (see Note 1).

1. Tare a piece of weighing paper.
2. Accurately weigh finely divided poly(styrene) into the weighing paper.
3. Carefully add the poly(styrene) to the volumetric flask.
4. Fill the flask two-thirds full with toluene.
5. Stopper and agitate gently.
6. Repeat agitation from time to time until the polymer is completely dissolved. Allow at least 24 hr of dissolution time.
7. Once the polymer is fully dissolved, fill the flask to the mark and mix well.
8. Filter the solution into a clean 100-ml Erlenmeyer flask and stopper it.
9. Rinse the filter with toluene.
10. To facilitate polymer dissolution, a magnetic heating and stirring unit may be used. Be careful to keep the temperature less than the boiling point of the solvent. Allow solution to return to room temperature before proceeding to step 7.

11. Check the viscometer for cleanliness while the polymer is dissolving. If the viscometer is visibly dirty, clean it with chromic acid (see Note 2).
12. Rinse viscometer thoroughly several times with water, then with acetone, and finally with toluene.
13. Dry the viscometer in an oven or pass clean, dry nitrogen or air through it.
14. Filter 200 ml of toluene to be used for cleaning the apparatus between runs and for analyses.

PROCEDURE

References 12, 13, and 36 provide further information on dilute solution viscosity experiments.

1. Using a 10-ml volumetric pipette, deliver 10 ml of filtered toluene to bulb A of the Ubbelohde viscometer (Fig. 17.1) (see Note 3).
2. Suspend the viscometer into the constant temperature bath and allow it to come to 25°C. This should take approximately 10–15 min. Record the temperature of the measurement.
3. Referring to Fig. 17.1, bring the solvent into bulb D by closing off tube 3 with one finger and applying pressure to tube 1 with either a rubber bulb or a stream of nitrogen.
4. When bulb D is partially full, unblock both tubes 1 and 3 and bulb B will drain, creating the suspended level at the bottom of the capillary in bulb B.

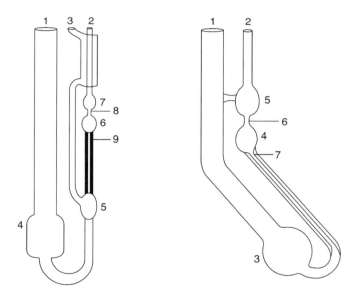

Fig. 17.1 Two types of viscometers: Ubbelohde (left) and Cannon-Fenske (right). The Ubbelodhe viscometer has the following components: (1) fill tube, (2) capillary outlet, (3) pressure relief tube, (4) solution bulb, (5) suspended volume bulb, (6) lower flow bulb, (7) upper flow bulb, (8) upper timing mark, and (9) lower timing mark. The Cannon-Fenske Viscometer has the following components: (1) fill tube, (2) capillary outlet tube, (3) solution bulb, (4) lower flow bulb, (5) upper flow bulb, (6) upper timing mark, and (7) lower timing mark.

5. Start timing when the bottom of the liquid meniscus (the concave surface of the liquid in the tube, see Fig. 17.2) passes the upper timing mark on the viscometer, E.
6. Time the flow until the bottom of the meniscus reaches the lower mark, F. This is time $= t_0$.
7. Repeat steps 3 to 6 at least three times. The readings should agree to within 0.1% of the average flow time (see Note 4).
8. Take the average of the readings for the solvent. This is time $= t_0$. Record all observations in Table 17.2.
9. Using a 5-ml volumetric pipette, add 5 ml of the filtered stock poly(styrene) solution to the solvent in bulb A.
10. Mix the resultant solution by closing off tube 3 and applying dry air or nitrogen to tube 2.
11. After thoroughly mixing the solution (5–10 min), repeat steps 2 to 8.
12. Using a 5-ml volumetric pipette, add another 5 ml of the filtered stock poly(styrene).
13. Repeat steps 2 to 11 until readings have been taken in four polymer solutions.
14. Thoroughly clean the viscometer as directed in the Procedure section.

FUNDAMENTAL EQUATIONS

$$\eta_r \simeq t/t_0$$
$$\eta_{sp} \simeq (t - t_0)/t_0$$
$$\eta_{red} = \eta_{sp}/c$$
$$\eta_{inh} = (\ln \eta_r)/c$$
$$\eta_{sp/c} = [\eta] + k'[\eta]^2 c$$
$$(\ln \eta_r)/c = [\eta] + k''[\eta]^2 c$$

CALCULATIONS

1. Prepare a table listing concentration (c), average flow time (t), η_r, η_{sp}, η_{red}, and η_{inh}, which are calculated from the equations. (*Note:* the concentration for the solvent, $c = 0$). Record the temperatures of the measurement and the calculated viscosities in Table 17.2.
2. Plot on the same piece of graph paper η_{red} versus c and η_{inh} versus c. Draw the best straight line you can through the set of points and extrapolate to $c = 0$. Read $[\eta]$ as the common intercept at $c = 0$ of the best straight line through the two sets of points (see Note 5).

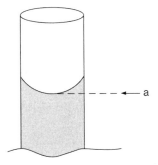

Fig. 17.2 Liquid meniscus in a tube: (a) meniscus bottom used in timing.

TABLE 17.2
Viscometric Parameters

c(g/100 ml)	Time	η_r	η_{sp}	η_{red}	η_{inh}

3. From the slope and the intercept of the η_{red} versus c plot, calculate the Huggins constant, k'.
4. From the slope and the intercept of the η_{inh} versus c plot, calculate the Kraemer constant, k''.
5. As an additional check on the accuracy of the experimental work, check to see that $k' - k'' = 0.5$.

REPORT

1. Describe the apparatus and experiment in your own words.
2. Include Table 17.2 and calculations.
3. Why does the intrinsic viscosity of a given polymer depend on the solvent employed?

NOTES

1. There are several ways to perform the dilution of the stock polymer solution.
   a. Start with the stock polymer solution and successively dilute with solvent. The efflux time of the solvent must be determined separately.
   b. Make up three or four separate polymer solutions to run. Determine the efflux time of the solvent. Rinse the viscometer with the next polymer solution to be run.
2. If another viscometer is being used, the operating instructions and representative figures should be followed.
3. Large variations in the flow time may result from foreign material in the capillary or from temperature variations. It may be necessary to reclean the viscometer.
4. Reliability in the intercept may be enhanced if the line is fitted by the method of least squares.

ACKNOWLEDGMENT

The authors thank Dr. Richard Perrinaud, Elf Atochem North America, for proofreading and making corrections.

REFERENCES

1. P. A. Lovell, in "Comprehensive Polymer Science: The Synthesis, Characterization, Reactions and Applications of Polymers" (G. Allen and J. C. Bevington, eds.), Vol. 1, Chap. 9, p. 173, Pergamon Press, New York, 1989.

2. P. J. Flory, "Principles of Polymer Chemistry," Cornell University Press, Ithaca, New York, 1953.
3. H. Tompa, "Polymer Solutions," Butterworths, London, 1956.
4. C. Tanford, "Physical Chemistry of Macromolecules," Wiley, New York, 1961.
5. H. Morawetz, "Macromolecules in Solution," 2nd Ed., Wiley, New York, 1975.
6. H.-G. Elias, "Macromolecules," Vol. 1, Wiley, New York, 1977.
7. N. C. Billingham, "Molar Mass Measurements in Polymer Science," Kogan Page, London, 1977.
8. P. F. Onyon, in "Techniques of Polymer Characterization" (P. W. Allen, ed.), p. 171, Butterworths, London, 1959.
9. W. R. Moore, in "Progress in Polymer Science" (A. D. Jenkins, ed.), Vol. 1, p. 1, Pergamon Press, Oxford, 1967.
10. J. B. Kinsinger, in "Encyclopedia of Polymer Science and Technology," Vol. 14, p. 717, Wiley, New York, 1971.
11. M. Bohdanecky and J. Kovár, "Viscosity of Polymer Solutions," Elsevier, Amsterdam, 1982.
12. E. M. Pearce, C. E. Wright, and B. K. Bordoloi, "Laboratory Experiments in Polymer Synthesis and Characterization," The Pennsylvania State University, University Park, PA, 1982.
13. E. A. Collins, J. Bareš, and F. W. Billmeyer, Jr., "Experiments in Polymer Science," Chap. 7C and Exp. 15, Wiley, New York, 1973.
14. M. L. Huggins, *J. Phys. Chem.* **42,** 911 (1938).
15. M. L. Huggins, *J. Phys. Chem.* **43,** 439 (1939).
16. P. Debye, *J. Chem. Phys.* **14,** 636 (1946).
17. J. J. Hermans, *Physica (Amsterdam)* **10,** 777 (1943).
18. H. A. Kramer, *J. Chem. Phys.* **14,** 415 (1946).
19. H. Mark, "Der Feste Köyser" (R. Saenger, ed.), p. 103 Hirzel, Leipzig, 1938.
20. R. Houwink, *J. Prakt. Chem.* **157,** 15 (1940).
21. W. Kuhn, *Kolloid-Z.* **68,** 2 (1934).
22. I. Sakurada, *Kasen-Koenshu* **5,** 33 (1940).
23. K. K. Chee, *J. Appl. Polym. Sci.* **30,** 1323 (1985).
24. D. Jadraque and J. M. Pereña, *Macromol. Chem.* **186,** 1263 (1985).
25. K. K. Chee, *Polymer* **28,** (1987).
26. A. R. Weiss and E. Cohn-Ginsberg, *J. Polym. Sci. Polym. Lett. Ed.* **7,** 379 (1969).
27. C. J. B. Dobbin, A. Rudin, and M. F. Tchir, *J. Appl. Polym. Sci.* **27,** 1081 (1982).
28. C. J. B. Dobbin, A. Rudin, and M. F. Tchir, *J. Appl. Polym. Sci.* **25,** 2985 (1980).
29. G. Meyerhoff, *Adv. Polym. Sci.* **3,** 59 (1961).
30. M. Kurata and W. H. Stockmeyer, *Adv. Polym. Sci.* **3,** 196 (1963).
31. J. Brandrup and E. H. Immergut, eds., "Polymer Handbook," 2nd Ed., Sects. IV.1 and IV.3, Wiley Interscience, New York, 1975.
32. I. Noda, T. Tsuge, and M. Nagasawa, *J. Phys. Chem.* **74,** 710 (1970).
33. A. Takahashi and M. Nagasawa, *J. Am. Chem. Soc.* **86,** 543 (1964).
34. J. A. Tan and S. P. Gasper, *J. Polym. Sci. Polym. Phys. Ed.* **13,** 1705 (1975).
35. ASTM D 2857-95 ("Standard Practice for Dilute Solution Viscosity of Polymers," 1995).

EXPERIMENT 18

Gel Permeation Chromatography

INTRODUCTION

The molecular weight (MW) and molecular weight distribution (MWD) are fundamental characteristics of a polymer sample. Gel permeation chromatography (GPC), more correctly termed size exclusion chromatography (SEC), is a separation method for polymers and provides a relative molecular weight [1–4]. Porath and Flodin [5] reported the first effective demonstration that polymers may be separated by the size dependence of the degree of solute penetration into a porous packing. The term gel permeation chromatography was defined by Moore [6], who developed rigid cross-linked polystyrene gels with a range of pore sizes, suitable for separation of synthetic polymers in organic media.

GPC is extremely valuable for both analytic and preparative work with a wide variety of systems ranging from low to very high molecular weights [2]. The method can be applied to a wide variety of solvents and polymers, depending on the type of gel used. GPC has been used for routine polymer characterization and quality control, particularly in determinations of MWD and for characterizing low polymers and small molecules, e.g., for prepolymers in resins and for polymer additives [7,8].

The use of multidetector GPC greatly increases the power of SEC, particularly in the case of copolymers. For copolymers of styrene-maleic anhydride (SMA), not only can the molecular weight distribution be determined, using a differential refractive index (DRI) detector, but also the compositional information of SMA (styrene content or acid number) by combining chromatograms from DRI and UV detectors [9].

THEORY

Background

All synthetic polymers show a distribution of molecular weights, which may be averaged in several ways. Any physical or performance property of a polymer may be related to one or more average molecular weights, the type of average is determined by the physical averaging process inherent in the method used to measure the property. Many polymers have distributions of other molecular parameters, such as chemical composition, stereoregularity, and chain branching.

Averages and distributions are well treated in many textbooks [10–17]. The aim here is to summarize the basic principles, essential definitions, and some examples of applications.

Size Exclusion

In the GPC experiment, polymer molecules are separated by size or their hydrodynamic volume because of their ability to penetrate part of the pores volume of the gel particles, i.e., the stationary phase. As the sample moves along the column with the mobile phase, the largest molecules are almost entirely excluded from the pores of the stationary phase, whereas the smallest find almost all the stationary phase accessible. The smaller the molecule, the more of the stationary phase volume is accessible to it and the longer it stays in that phase. Consequently, small molecules are eluted from the column later (Fig. 18.1).

The separation of a solute of a given size in solution is determined by a distribution coefficient, K_{sec}, which governs the fraction of internal pore volume of the gel, V_i, that is accessible to this solute. The value of the retention volume, V_r, for this solute is given by

$$V_r = V_0 + K_{sec} V_i, \qquad (1)$$

where V_0 is the interstitial or mobile phase volume and K_{sec} is the ratio of pore volume accessible to a species to the total pore volume. For very large molecules, completely excluded from the gel, V_r is equal to V_0. Very small molecules have free access to both stationary and mobile phases, i.e., K_{sec} is unity. For intermediate species, K_{sec} is a separation constant between 0 and 1. The dependence of

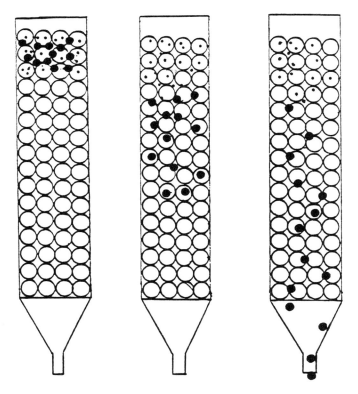

Fig. 18.1 GPC column separation of a polymer.

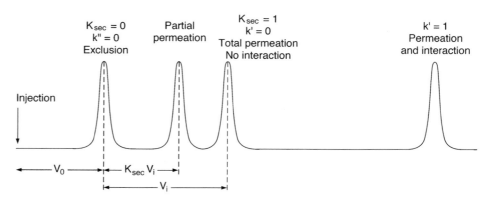

Fig. 18.2 Elution of peaks in gel permeation chromatography.

V_r on solute is shown in Fig. 18.2 for solutes with a range of sizes, e.g., for calibration standards.

Benoit and co-workers [18] proposed that the hydrodynamic volume, V_η, which is proportional to the product of $[\eta]$ and M, where $[\eta]$ is the intrinsic viscosity of the polymer in the SEC eluent, may be used as the universal calibration parameter (Fig. 18.3). For linear polymers, interpretation in terms of molecular weight is straightforward. If the Mark–Houwink–Sakurada constants K and α are known, $\log [\eta]M$ can be written $\log M^{1+\alpha} + \log K$, and V_r can be directly related to M. The size-average molecular weight, $\overline{M}_z$, is defined by this process:

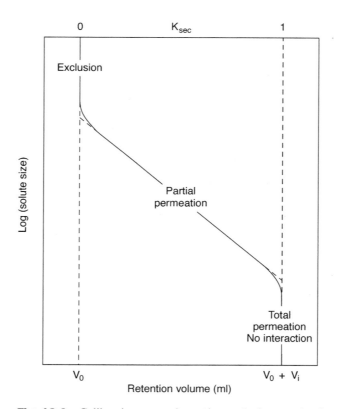

Fig. 18.3 Calibration curve for a size exclusion mechanism.

$$\overline{M}_z = \frac{\sum\limits_{i=1}^{\infty} w_i M_i^{1+\alpha}}{\sum\limits_{i=1}^{\infty} w_i M_i}. \qquad (2)$$

For branched polymers, molecular size is crucial because the material eluting at any value of V_r consists of a mixture of species having different molecular weights and degrees of branching but constant hydrodynamic volume.

Column Efficiency

Plate Height and Plate Number

A measure of the efficiency of the chromatography column is the height equivalent to a theoretical plate or plate height H [19]. The plate height for an experimental chromatogram is calculated from

$$H = L/N, \qquad (3)$$

where L is the column length and N is the plate number. If the peaks in the chromatograms are symmetrical (Fig. 18.4), corresponding to a normal error (or Gaussian) function, then N may be determined from

$$N = 16 \, (V_e/W)^2, \qquad (4)$$

where W is the line width at half-height of the chromatogram and V_e is the eluting volume for the solute ($V_e = V_o + V_i$). A typical microparticulate packing with a particle diameter of $\sim 10 \ \mu m$ will generate a high-performance (HP) SEC column having $N > 20{,}000$ plates m^{-1}.

Column Packings

Essential conditions for the effective fractionation of polymers by SEC are that the pore sizes in the column packing should be comparable to polymer sizes in solution and that the packings should have substantial pore volume, typically $0.5 < V_i/V_o < 1.65$ for macroporous packing. Consequently, to separate samples with a wide range of molecular weights it is necessary to have a series arrangement of columns, each covering a different molecular size range or to use a single column containing several gels having various pore size distributions. It is advantageous for the pore size distribution to generate a linear relation between log M and V_R in order to facilitate molecular weight analysis.

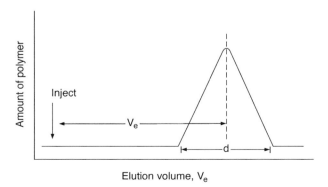

Fig. 18.4 Calculation of plate count in GPC.

Rigid microparticulate packings are generally employed in HPSEC [7,20–22]. Packings with particle diameters of ~10 μm are typically used for high polymers, with high resolution separations of low polymers, prepolymers, and small molecules being performed with particles having diameters of ~3 μm [17]. Separations of long chain polymers with microparticulate packings must be performed carefully in order to avoid shear degradation during macromolecular diffusion through the column.

Eluents

Ideally, the eluent should be a good solvent for the polymer, should permit high detector response from the polymer and should wet the packing surface [17]. The most common eluents in SEC are tetrahydrofuran for polymers that dissolve at room temperature, o-dichlorobenzene and trichlorobenzene at 130–150°C for crystalline polyalkenes and m-cresol and o-chlorophenol at 90°C for crystalline condensation polymers such as polyamides and polyesters. For more polar polymers, dimethylformamide and aqueous eluents may be used, however, care is required in avoiding solute–gel interaction effects. Secondary retention mechanisms are always likely to occur when polymer–solvent interactions are not favorable, when polar polymers are separated with less polar eluents, and when packings have active surface sites. Therefore, it may be necessary to carefully select the eluent composition, e.g., by the addition of an electrolyte or organic component, after assessing the nature of possible polymer-packing interactions [23,24].

Detectors

In SEC the concentration by weight of polymer in the eluting solvent may be monitored continuously with a detector measuring refractive index, UV absorption, or infrared (IR) absorption [17]. The resulting chromatogram is therefore a weight distribution of the polymer as a function of retention volume, V_R.

Experimental SEC conditions require highly sensitive concentration detectors giving a detector response, which is linearly related to polymer concentration. The most common detector for monitoring polymer concentration in the eluent is the differential refractometer (DRI). The response of the detector to polymer concentration does not depend on polymer molecular weight except for very low polymers.

The most sensitive detector is the differential UV photometer, which is appropriate for a polymer with a significant UV absorbance with a nonabsorbing eluent. This detector is not appreciably affected by flow pulsations, flow rate changes, and temperature fluctuations.

When characterizing copolymers, it is necessary to have two detectors in series, e.g., a refractometer with either a UV detector or an IR detector. An IR detector is preferred for the detection of polyalkenes at elevated temperatures because baseline noise and drift are much less than for the refractometer detector.

If the universal calibration is valid, then at a given V_R the following relationship will apply:

$$\log[\eta]_p M_p = \log[\eta]_{ps} M_{ps}, \qquad (5)$$

where p refers to a polymer requiring characterization and ps to polymer standards, which will be polystyrene or poly(methyl methacrylate) for organic eluents and poly(ethylene oxide) and/or polysaccharide standards for aqueous eluents [17]. $[\eta]$ may be measured using an on-line viscometer detector. The calibration curve M_p may be determined when the dependence of $[\eta]_p$ and $[\eta]_{ps} \cdot M_{ps}$ on V_R has been established.

When the molecular weight of the polymer in the eluting solvent is measured experimentally with a low-angle, laser light-scattering detector (LALLS), then the dependence of w(M) on M can be established directly (Table 18.1). In LALLS, the intensity of scattering from the polymer is expressed in terms of the excess Rayleigh factor, R_θ, defined as the scattering intensity of the polymer solution minus the scattering intensity of the solvent at a given angle θ normalized with respect to the intensity of the incident beam and the scattering volume. The value of R_θ will be a function of the scattering angle, the polymer concentration, and the polymer molecular weight. Although light scattering gives good scattering intensities for polymers having high M, there may be little or no LALLS detector sensitivity with $M < 10^4$.

For a polydisperse polymer, experimental measurements of M for the chromatogram at high V_R may not be accurate. When average molecular weights are computed from the distribution w(M) derived from data obtained with concentration and molecular weight detectors, the value of $\overline{M}_w$ is likely to be more valuable than $\overline{M}_n$, which could be substantially in error [25,26].

APPLICABILITY

GPC techniques are applicable to a wide variety of solute materials, both low and high molecular weight, dissolved in solvents of varying polarity. The selection of column type, column pore size, solvent, and temperature must be appropriately made for each solute. Care must be taken to avoid reaction between the solute and the columns or other adsorption phenomena, especially when two solvents are used—one to dissolve the solute and one in the chromatograph. Changing from one solvent to another in the chromatograph can take 24 hr before the baseline stabilizes.

The effect of the solvent on the hydrodynamic volume of the polymer must be understood as GPC is not an absolute technique but needs to be calibrated versus standards. The direct interpretation of GPC results in terms of the molecu-

TABLE 18.1
Summary of the Molecular Weight Averages Most Widely Encountered in Polymer Chemistry

Average	Definition	Alternative form	Absolute methods of measurement
Number, $\overline{M}_n$	$\dfrac{\sum_i N_i M_i}{\sum_i N_i}$	$\dfrac{\sum_i W_i}{\sum_i (W_i/M_i)}$	Osmotic pressure and other colligative properties. End group analysis
Weight, $\overline{M}_w$	$\dfrac{\sum_i N_i M_i^2}{\sum_i N_i M_i}$	$\dfrac{\sum_i W_i M_i}{\sum W_i}$	Light scattering, sedimentation velocity
Size, $\overline{M}_z$	$\dfrac{\sum_i N_i M_i^3}{\sum_i N_i M_i^2}$	$\dfrac{\sum_i W_i M_i^2}{\sum_i W_i M_i}$	Sedimentation equilibrium
Viscosity, $\overline{M}_v$	$\left(\dfrac{\sum_i N_i M_i^{1+a}}{\sum_i N_i M_i}\right)^{1/a}$		Intrinsic viscosity (see Chapter 17)

lar weight distribution is applicable only to linear homopolymers where the calibrating materials and the test samples are of the same chemical type.

Use of the differential refractometer detector is applicable to all polymers having refractive indices different from that of the solvent. However, a correction must be made if the polymer refractive index depends on molecular size, such as at very low molecular weights.

ACCURACY AND PRECISION

GPC is a relative method, especially for UV and refractive index (RI) detectors, and must be calibrated using polymer standards whose molecular weight has been determined using absolute methods such as intrinsic viscosity or light scattering. Consequently, the accuracy is relative to the calibration.

Repeat chromatograms in GPC should agree within 2%. However, the chromatogram is sensitive to such experimental conditions as (a) resolution of the columns, (b) range of porosities of the column packings, (c) flow rate of solvent, and (d) age of detectors. It is recommended that the laboratory use "control charts" to determine the optimal conditions of the instrument.

SAFETY PRECAUTIONS

Safety glasses must be worn in the laboratory at all times. Material safety and data sheets should be read prior to the start of the experiment. All chemicals should be considered hazardous from a standpoint of flammability and toxicity. Appropriate safety gloves must be worn when using organic solvents so that no skin contact is permitted. Care must be taken to use organic solvents either in a well-ventilated area or in a hood. Avoid breathing the fumes or sources of electrical sparks. The GPC instrument, including solvent reservoir and waste container, should be vented to a fume hood or other exhaust system.

APPARATUS

1. Gel permeation chromatograph with variable temperature accessory
2. Appropriate detectors for the analyses, preferably a variable wavelength UV detector as well as a refractive index detector
3. Appropriate column set suitable for the analysis of high molecular weight (see Notes 1 and 2)
4. 25-ml stoppered volumetric flasks
5. Fine-fritted glass or 0.45-μm Millipore filter in a suitable holder
6. Hypodermic syringe appropriate to deliver samples into the GPC
7. Pipettes for making up polymer solutions
8. Rubber bulb for the pipette
9. Sample containers and stoppers for all samples
10. Laboratory balance capable of weighing to 0.0001 g
11. Extra column prefilters
12. Aluminum weighing dishes or weighing paper.

REAGENTS AND MATERIALS

1. Polydispersed poly(styrene) (see Note 1)
2. Poly(styrene) standards, anionic, narrow distribution, with $\overline{M}_w$ covering the range of 10,000 to 1,000,000

3. Tetrahydrofuran (THF), either distilled reagent grade or high-pressure liquid chromatography (HPLC) grade (see Note 1)
4. o-Dichlorobenzene, reagent grade
5. Sulfur
6. High purity dry nitrogen gas.

PREPARATION

Prepare the polymer solutions in volumetric flasks ahead of time with concentrations for the poly(styrene) narrow distribution standards at 0.8g/liter and the polydisperse poly(styrene) at 1.2g/liter.

1. Tare a clean, dry volumetric flask.
2. The high-purity sulfur, the flow rate standard, should be made up in sufficient quantity, such that it may be used as solvent for the polymer standards and the polymer unknown sample (see Note 3)
   a. Weigh a piece of weighing paper or aluminum weighing dish. Zero tare.
   b. Weigh 5 mg of sulfur powder.
   c. Record the weight of the sulfur.
   d. Place the sulfur into a 50-ml volumetric flask.
   e. Add approximately 35 ml of THF to the volumetric flask and cap.
   f. Allow several hours to dissolve, gently agitating occasionally.
   g. Once dissolved, add remaining THF to the level of the meniscus of the volumetric flask and cap.
   h. Invert solutions several times carefully to ensure a homogeneous solution.

3. The polymer sample to be analyzed is treated in the following way:
   a. Tare a clean, dry 10-ml volumetric flask.
   b. Accurately weigh 10 mg of the polymer into the volumetric flask. Record the weight.
   c. Rinse the 4-ml pipette twice with pure THF and blow it dry with high purity dry nitrogen gas.
   d. Pipette 4-ml of the sulfur/THF solution into a volumetric flask.
   e. Add another 2–3 ml pure THF to the volumetric flask and cap.
   f. Allow several hours to dissolve (generally 1–2 hr), agitating gently.
   g. Once dissolved, add remaining THF to the level of the meniscus of the volumetric flask and cap.
   h. Invert the solutions several times carefully to ensure a homogeneous solution (see Note 4).
   i. Filter the solutions before injection into the chromatograph (see Note 5).

4. The narrow MWD polymer standard solutions are made up with two or more polymers in the same flask (see Notes 1 and 6).
   a. Tare a clean, dry 10-ml volumetric flask.
   b. Accurately weigh 10 mg of each of the narrow MWD poly(styrene) standards into the volumetric flask.
   c. Record the weight of the polymer.
   d. Rinse the 4-ml pipette twice with pure THF and blow it dry with high purity dry nitrogen gas.
   e. Pipette 4 ml of the sulfur/THF solution into the volumetric flask.
   f. Add another 2–3 ml pure THF to the volumetric flask and cap.
   g. Allow several hours to dissolve (generally 1–2 hr), agitating gently.

h. Once dissolved, add the remaining THF to the level of the meniscus of the volumetric flask and cap.
   i. Invert the volumetric several times carefully to ensure a homogeneous solution (see Note 4).
   j. Filter the solutions before injection into the chromatograph (see Note 5).

5. Filtering chemical plant samples
   a. Rinse a 5-ml syringe and plunger twice with pure THF.
   b. Screw a 0.45-μm PTFE filter onto the tip of the syringe.
   c. Label and uncap a clean GPC vial.
   d. Place syringe over clean vial.
   e. Pour sample solution into syringe.
   f. Insert plunger and press until liquid passes through the filter into a clean vial.
   g. Immediately cap filtered sample.
   h. Discard used vial, cap, and filter into an appropriate waste container.
   i. Repeat steps a–h for each sample.
   j. Rinse syringe twice with pure THF when finished filtering samples.

PROCEDURE

Instrument Setup

Due to the large number of instrument types available on the market, no attempt is made to write a specific procedure. It is recommended that the instrument manual be consulted and Refs. 27 to 31 be read prior to starting this experiment. The experiment will take at least 3 hr.

   a. The instrument should be stable with a flat baseline. Short-term noise should not exceed 2% of the maximum.
   b. Columns should be conditioned with the solvent used for analysis for at least 24 hr.
   c. The solvent reservoir should be degassed [13] prior to and throughout the run.
   d. Be sure to check the prefilter to the column set prior to the analysis. If it is suspected to be plugged or damaged, change the prefilter.
   e. Be sure that the temperature of the GPC systems, especially that of the detectors, has stabilized at 40°C.
   f. Detector settings should be selected to optimize the detector response.
   g. Record the temperature of each of the components before and after the run. (Temperature fluctuation will affect the separation.)
   h. The flow rate of the GPC system is determined by the column and instrument manufacturers [29–31]. A flow rate of 1 ± 0.1 ml $\cdot$ min^{-1} is suggested [29–31] but not required. The flow rate needs to be kept constant during the calibration and measurement.

FUNDAMENTAL EQUATIONS

$$\overline{M}_n = \frac{\sum_i n_i M_i}{\sum_i n_i} = \frac{\sum_i (h_i/M_i)(M_i)}{\sum_i (h_i/M_i)} = \frac{\sum_i h_i}{\sum_i (h_i/M_i)}$$

$$\overline{M}_w = \frac{\sum_i w_i M_i}{\sum_i w_i} = \frac{\sum_i h_i M_i}{\sum_i h_i}$$

CALCULATIONS

A number of calculation programs are available to automate the following.

1. Read the retention volume, V_r, for each of the narrow distribution polymer standards.
2. Using the retention volume, V_r, for the narrow distribution polymers and their molecular weight, M, a calibration curve is constructed for log M versus V_e. $\overline{M}_w$ may be used for M.
3. Data from the GPC of the unknown polymer samples are used to calculate the $\overline{M}_n$, $\overline{M}_w$, and $\overline{M}_w/\overline{M}_n$, and molecular weight distribution curves. The amount of polymer at any particular V_e or corresponding molecular weight, through correlation using the calibration curve, is proportional to the height of the curve, H_i, at that particular V_e:

$$H_i \propto n_i M_i,$$

where n_i is the moles of the ith polymer and M_i is the molecular weight of that polymer. Data can most easily be handled in tabular form (Table 18.2).

4. The $\overline{M}_n$ and $\overline{M}_w$ of the samples can be calculated from data in Table 18.2 using Equations. (1) and (3). In these calculations, the more divisions for V_e, the better will be the accuracy of the figures derived. After $\overline{M}_w$ and $\overline{M}_n$ are found, the polydispersity $\overline{M}_w/\overline{M}_n$ is calculated.
5. A molecular weight distribution curve, which is a plot of the amount of polymer ($w_i = n_i w_i$) in arbitrary units (e.g., $H_i \propto n_i w_i$) vs M_i, can be obtained from data in Table 18.2.
6. Column efficiency can be calculated from the following equation:

$$\text{Plate count}, N = \frac{16}{f}\left(\frac{V_e}{d}\right)^2 \text{ or } 16\left(\frac{V_e}{W}\right)^2,$$

where N is the number of theoretical plates per foot (f) of column, and V_e and d are the elution volume (or time) measured at the peak maximum and peak base in elution volume (or time) units as determined by measuring the distance between the baseline intercept of lines drawn tangent to the peak inflection points, as shown in Fig. 18.4.

REPORT

1. Describe the experiment and apparatus in your own words.
2. Include all plots and tables, the chromatograms, and the calculated values of $\overline{M}_n$, $\overline{M}_w$ and $\overline{M}_w/\overline{M}_n$ and the plates per foot.
3. Comment on the resolution of the chromatograph in relation to (a) the plate count and (b) to separation between the members of a pair of narrow distribution samples.

TABLE 18.2
GPC Parameters for Calculation of Molecular Weight Averages and Distribution

V_r	M_i	$H_i \propto n_i M_i$	n_i	$n_i M_i^2$

4. What changes in procedure or interpretation would be required in order to analyze the following by GPC:

   a. Poly(α-methylstyrene)
   b. Branched poly(styrene)
   c. Poly(styrene-co-methylmethacrylate).
   d. What kind of information could be obtained?

5. How would you prepare a calibration curve if you were operating in a solvent in which poly(styrene) is not soluble?

NOTES

1. Polymer standards should be unimodal, narrow MWD ($\overline{M}_w/\overline{M}_n < 1.1$) of known molecular weight (Fig. 18.5).

   a. Polymer Lab produces prepackaged poly(styrene) standards, EasyCal PS-2, in low and high molecular weight ranges. These standards, once made up, can be reused so they do not have to be prepared for every run.

2. Low molecular weight compounds, such as o-dichlorobenzene or sulfur, that are used for determining plate count or as internal standards must be of high purity.
3. Standard and polymer concentrations will need to be adjusted depending on the age of the instruments and columns and the sensitivity of the detectors.
4. The polymer should be dissolved at room temperature [28]. Magnetic stirring devices or laboratory shakers are recommended to aid dissolution. Excessive temperature or ultrasonic devices may cause the polymer to degrade. Polystyrene solutions prepared with solvents such as THF are very stable, as long as MW < 500,000 g mol^{-1}. However, it is a good practice to analyze polymer solutions within 24 hr of their preparation [28].

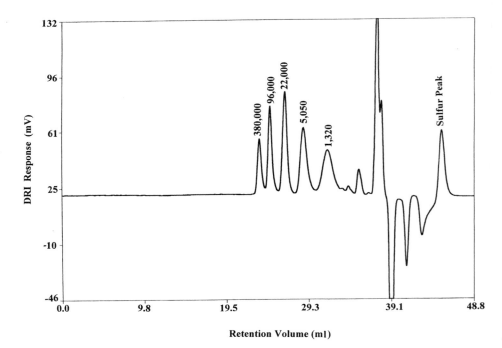

Fig. 18.5 Poly(styrene) high molecular weight standards.

5. It is recommended that all solutions be filtered through membrane filters to remove lint, polymer gel, and other materials likely to obstruct the columns and other system components.

ACKNOWLEDGMENTS

The authors thank Dr. Richard Perrinaud, Elf Atochem North America, for proofreading and providing some figures.

REFERENCES

1. J. V. Dawkins, in "Comprehensive Polymer Science: The Synthesis, Characterization, Reactions and Applications of Polymers: Polymer Characterization" (G. Allen and J. C. Bevington, eds.), Vol. 1, Chap. 12, p. 231, Pergamon Press, New York, 1989.
2. F. W. Billmeyer, Jr., "Textbook of Polymer Science," 3rd Ed., Wiley, New York, 1984.
3. I. Tomka and G. Vaneso, "Applied Polymer Analysis and Characterization: Recent Developments in Technique, Instrumentation, Problem Solving" (J. Mitchell, Jr., ed.), Chap. 11-J, p. 237. Hanser Publishers, New York, 1987.
4. P. C. Hiemenz, "Polymer Chemistry: The Basic Concepts," Dekker, New York, 1984.
5. J. Porath and P. Flodin, *Nature (London)* **183,** 1657 (1959).
6. J. C. Moore, *J. Polym. Sci. A* **2,** 835 (1964).
7. W. W. Yau, J. J. Kirkland, and D. D. Bly, "Modern Size Exclusion Liquid Chromatography, Practice of Gel Permeation and Gel Filtration Chromatography," Wiley, New York, 1979.
8. J. Janca, "Steric Exclusion Liquid Chromatography of Polymers," Dekker, New York, 1984.
9. R. Amin-Sanayei, private communication.
10. H.-G. Elias, "Macromolecules," 2nd Ed., Plenum, New York, 1984.
11. G. Odian, "Principles of Polymerization," 2nd Ed., Wiley-Interscience, New York, 1981.
12. P. J. Flory, "Principles of Polymer Chemistry," Cornell, Ithaca, NY, 1953.
13. H.-G. Elias, R. Bareiss, and J. G. Watterson, *Adv. Polym. Sci.* **11,** 111 (1973).
14. H.-G. Elias, *Pure Appl. Chem.* **43,** 115 (1975).
15. A.-Q. Tang, "Statistical Theory of Polymer Reactions," Academic Press, Beijing, 1985.
16. L. H. Peeples, Jr., "Molecular Weight Distributions in Polymers," Wiley–Interscience, New York, 1971.
17. C. Booth and R. O. Colclough, in "Comprehensive Polymer Science: The Synthesis, Characterization, Reactions and Applications of Polymers: Polymer Characterization;" eds. G. Allen and J. C. Bevington, Volume 1, Chapter 3, p.55, Pergamon Press, New York, 1989.
18. Z. Grubisic, P. Rempp and H. Benoit, *J. Polym. Sci., Polym. Lett. Ed.* **72,** 753 (1967).
19. J. C. Giddings, "Dynamics of Chromatography, Part 1, Principles and Theory," Dekker, New York, 1965.
20. R. E. Majors, *J. Chromatogr. Sci.* **15,** 334 (1977), **18,** 488 (1980).
21. K. K. Unger and J. N. Kinkel, "Aqueous Size Exclusion Chromatography," J. Chromatogr. Library, ed. P. L. Dubin, Elsevier, Amsterdam, 1988, Vol. 40.
22. B. G. Belenkii and L. Z. Vilenchik, "Modern Liquid Chromatography of Macromolecules," J. Chromatogr. Library, Vol. 25, Elsevier, Amsterdam, 1983.
23. P. L. Dubin, *J. Liq. Chromatogr.* **3,** 623 (1980).
24. G. Coppola, P. Fabbri, B. Pallesi and U. Bianchi, *J. Appl. Polym. Sci.* **16,** 2829 (1972).
25. R. C. Jordan, *J. Liq. Chromatogr.* **3,** 439 (1980).
26. R. C. Jordan and M. L. McConnell, *ACS Symp. Ser.* **138,** 107 (1980).
27. E. M. Pearce, C. E. Wright and B. K. Bordoloi, "Laboratory Experiments in Polymer Synthesis and Characterization," The Pennsylvania State University, University Park, PA, 1982.
28. ASTM D 5296-92.
29. ASTM D 3016.
30. ASTM D 3536-91.

EXPERIMENT 19

Light Scattering

INTRODUCTION

Light scattering occurs whenever a beam of light encounters matter. When there is no absorption, nuclei and electrons undergo induced vibrations in phase with the incident light wave and act as sources of light that is propagated in all directions, aside from a polarization effect with the same wavelength as the exciting beam. Light scattering accounts for many natural phenomena, including the colors of the sky and the rainbow.

The scattering of light has interested both scientists and laymen for many years. Lord Rayleigh [1,2] developed the theoretical interpretation of light scattering from dilute gases. Einstein [3] and Smoluchowski [4,5] explained scattering in liquid on the basis of local thermal fluctuations in density in the medium. Working from this basis, Debye [6-8] extended the work to macromolecular solutions and showed a relationship between local fluctuations and osmotic pressure.

Light scattering from solution allows the determination of the molecular parameters [9-18] (molecular weight, dimensions, shapes, etc.) of the scattering particles and thermodynamic quantities (virial coefficients, chemical potential, preferential adsorption coefficients, and excess free energies of mixing) [15-28]. Because of the importance of light scattering, many review articles have been published [9-11,14,24-36].

THEORY

Background

If the frequency of the scattered radiation is the same as that of the incident radiation (i.e., they have the same energy), the scattering is "elastic." If the scattering process involves an energy exchange between the radiation and the scattering particles, the scattering is "nonelastic," e.g., Raman and Brillouin scattering [37].

In the case of solutions, concentration fluctuations only contribute to the central elastic part of the scattering spectrum. However, the Brownian movement of solute molecules creates weak frequency displacements that broaden the central peak. This phenomenon is called "Rayleigh line broadening" or "quasielastic" scattering [26-28]. This section deals with elastic scattering only.

Light Scattering from a Liquid

Light scattering from a solution is due both to the scattering from local density fluctuations and to the scattering from the solvent [9,18]. This scattering may be described by the Rayleigh scattering ratio [9,18]:

$$R_\theta = (I_\theta/I_0) \cdot r, \tag{1}$$

where I_0 is the intensity of the incident light, I_θ is the intensity of the scattered light at the angle, θ, and r is the distance of the detector from the scattering sample. As I_0 and r can be determined from the optical and mechanical characteristics of the photometer used, the Rayleigh ratio can be obtained experimentally.

Scattering from a Solution of Small Particles

In solutions and in mixtures of liquids, additional light scattering arises from irregular changes in density and refractive index due to fluctuations in composition. If the solution is dilute, the density fluctuations are essentially identical to those existing in the pure solvent [9]

$$\Delta R_\theta = R_{\theta,\, \text{solution}} - R_{\theta,\, \text{solvent}} = \frac{\pi^2 \delta V}{2\lambda_0^4} \cdot \left(\frac{\partial \varepsilon}{\partial c}\right)^2 \cdot (1 + \cos^2\theta) \cdot \overline{\Delta c^2}, \tag{2}$$

where λ_0 is the wavelength of incident light, δV is the density fluctuations, and $\partial \varepsilon / \partial c$ is the change in dielectric constant with concentration [9]. Following Maxwell's relation, $\varepsilon = n^2$, where n is the refractive index [9],

$$\left(\frac{\partial \varepsilon}{\partial c}\right)^2 = 4n_1^2 \left(\frac{dn}{dc}\right)^2, \tag{3}$$

where n_1 is the solvent refractive index and (dn/dc) is the differential changes in the refractive index as a function of solute concentration [9,10].

From thermodynamic arguments, $\overline{\Delta c^2}$ is related to the chemical potential, μ. Thus, the excess Rayleigh ratio may be written [9]

$$\Delta R_\theta = \frac{2\pi^2 \, n_1^2 \, kT \, (dn/dc)^2 \, \overline{V}_1 \, c}{\lambda_0^4 \, (\partial \mu_1 / \partial c)_T} \cdot (1 + \cos^2\theta), \tag{4}$$

where $\overline{V}_1$ is the partial molar volume.

The relation between the chemical potential of the solvent and the osmotic pressure Π is

$$\Pi \overline{V}_1 = -(\mu_1 - \mu_1^0), \tag{5}$$

where μ_1^0 is the chemical potential of the pure solvent [9]. The excess Rayleigh ratio can be rewritten

$$\Delta R_\theta = \left[\frac{K \, R \, Tc}{(\partial \Pi / \partial c)}\right] \cdot (1 + \cos^2\theta), \tag{6}$$

where

$$K = \frac{2\pi^2 n_1^2}{\lambda_0^4 N_A} - (dn/dc)^2. \tag{7}$$

K is also known as the Debye constant [9], N_A is Avogadro's number, and n_0 is the refractive index of the pure solvent.

The osmotic pressure may be written as a power series in concentration [9].

$$\frac{\Pi}{c} = RT \left(\frac{1}{M} + A_2 c + A_3 c^2 + \ldots\right), \tag{8}$$

where M is the average molecular weight. By rearranging, the general equation for Rayleigh scattering of unpolarized light from a real solution of particles is much smaller than the wavelength of light (i.e., less than 1/20th of the wavelength of light) becomes [9]:

$$\frac{Kc(1+\cos^2\theta)}{\Delta R_\theta} = \frac{1}{M} + 2A_2c + 3A_3c^2 + \ldots, \tag{9}$$

where $A_2, A_3, \ldots$ are the second, third, $\ldots$ virial coefficients, M is the average molecular weight, c is the polymer concentration, and ΔR_θ is the difference in scattering between the solution and the pure solvent.

Scattering from a Solution of Larger Particles

When the size of a scattering particle exceeds $\lambda/20$, different parts of the particle are exposed to incident light of different amplitude and phase [9,16,18]. The scattered light is made up of waves coming from different parts of the particle that interfere with one another [9,16,18]. Consequently, the scattered light intensity varies with the angle. Debye and others described this variation with a *particle scattering factor* $P(\theta)$, which depends on the model selected to describe the scattering system, such as a sphere, random coil, or other type [9,16,18].

$$P(\theta) = 1 - \frac{16\pi^2}{3\lambda_s^2} \langle R_g^2 \rangle \sin^2(\theta/2), \tag{10}$$

where $\langle R_g^2 \rangle$ is the mean square radius of gyration and λ_s is the wavelength of light in the solution: $\lambda_s = \lambda_0/n$. The relationship between $P(\theta)$ and the mean-square radius of gyration of a macromolecule is a useful general result, as it is independent of molecule shape. The limiting slope at $\theta = 0$ of a plot of $P(\theta)$ against $\sin^2(\theta/2)$ for a macromolecule is $-16\pi^2 \langle R_g^2 \rangle/3\lambda^2$, which permits determination of $\langle R_g^2 \rangle$ [9].

Treatment of Data: General Equation and Zimm Plot

When the particles are large enough to display angular dissymmetry, Eq. (9) is only valid when $\theta = 0$, as $P(\theta) = 1$ for the angle. The experimental scattering intensities can only be analyzed as a function of the observation angle as it is impossible to measure the scattered intensity at zero angle [9]

$$\frac{Kc(1+\cos^2\theta)}{\Delta R_\theta} = \frac{1}{MP(\theta)} + 2A_2Q(\theta)c, \tag{11}$$

where the factor $Q(\theta)$ is due to intermolecular interference effects at finite concentrations. $Q(\theta)$ falls below unity only for scattering from large particles (high molecular weight polymers in good solvents) at large values of θ. For $c \to 0$ and $\theta \to 0$, the expression for the scattering intensity for unpolarized light becomes (with mathematical manipulation) [9]

$$\frac{Kc(1+\cos^2\theta)}{\Delta R_\theta}\Big|_{\substack{c\to 0 \\ \theta \to 0}} = \frac{1}{M}\left(1 + \frac{16\pi_2}{3\lambda_2}\langle R_g^2 \rangle \frac{\sin^2\theta}{2}\right) + 2A_2c. \tag{12}$$

This equation implies a double dependence of scattering intensities on concentration and observation angle [9]. By extrapolating the scattering data for each concentration to zero angle, the second virial coefficient, which is related to thermodynamic properties, may be measured [9,10,15–18].

$$\left.\frac{Kc\,(1+\cos^2\theta)}{\Delta R_\theta}\right|_{\theta=0} = \frac{1}{M} + 2A_2 c. \tag{13}$$

By extrapolating scattering data for each angle to zero concentration, the mean-square radius of gyration may be measured [9,10,15–18]

$$\left.\frac{Kc\,(1+\cos^2\theta)}{\Delta R_\theta}\right|_{c=0} = \frac{1}{M} + \frac{16\pi^2}{3\lambda^2 M}\langle R_g^2\rangle \sin^2(\theta/2). \tag{14}$$

The common intercept of the two plots gives the reciprocal of the molecular weight. Zimm [38–40] proposed a graphical method to do this double extrapolation. This method [9,10,15–18,38–41] consists of plotting the value of $Kc\,(1+\cos^2\theta)/\Delta R_\theta$ for each concentration and angle against $\sin^2(\theta/2) + k'c$. The constant k' is arbitrary and is used to allow a separation of data points. For high molecular weights, a practical method is to choose a value of k' that gives a value near unity when multiplied by the smallest concentration. For low molecular weights, the highest concentration is used. An example of a Zimm plot is shown in Fig. 19.1.

Instrumentation

Calibration

Direct measurements of absolute values of scattered light intensity are experimentally rather difficult. A comparative measurement is often performed. The intensity of light scattered by the sample is compared with the scattering intensity of a standard under similar conditions. In practice, the quantities that are compared are the detector intensities recorded with the sample and standard, respectively, from which the scattering intensity of the sample, expressed in terms of the Rayleigh ratio, R_θ, is calculated [10,41]. The calibration standards are pure liquids, solutions of some standard polymers, or dispersions of standard colloids. The use of pure liquids as calibration standards has the advantage that their scattering power depends only on the temperature and the wavelength of incident light [10,41]. A standard liquid must fulfill several requirements: (1) a large scattering intensity, (2) easy purification, and (3) stable under UV light. Benzene is widely used as a calibration standard [9,10] (*Caution:* Benzene is highly toxic and a cancer suspect agent.) The Rayleigh ratio for the standard is given by [9]

$$R_{S,90} = I_S\,(90°)\,r^2/I_0. \tag{15}$$

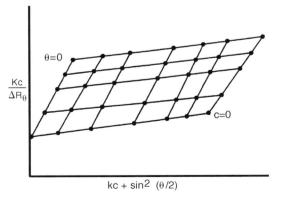

Fig. 19.1 Zimm plot for an arbitrary polymer showing the double intercept.

The experimental measurements may be expressed as [9]

$$\Delta R_\theta = R_{S,90} (n_0^2/n_S^2) (\Delta I_\theta/I_{S,90}) \sin\theta. \quad (16)$$

The ratio (n_0^2/n_S^2) is the refractive index correction to the scattering volume [42]. Equation (12) becomes the following expression [9]:

$$k'c (I_{S,90}/\alpha\Delta I_\theta) (dn/dc)^2 = (1/M) [1 + (16\pi^2/3\lambda^2) \langle R_g^2 \rangle \sin(\theta/2)] + 2A_2c, \quad (17)$$

where $\alpha = \sin\theta/(1 + \cos^2\theta)$ and $k' = (2\pi^2 n_S^2/\lambda_0^4 N R_{S,90})$.

Correction Factors

The observed ratio of detector intensities has to be corrected by a few factors in order to obtain the two Rayleigh ratios [10,15–18,41,42]. Typical phenomena that must be corrected for are [10,41]: (1) scattering volume, (2) refraction correction, (3) reflection correction, (4) absorption, (5) fluorescence, (6) polarization, and (7) sensitivity of photomultiplier.

Clarification

Solvents and solutions used for light scattering measurements should be absolutely dust free. Dust particles disturb the angular distribution of the scattering intensities, especially at low angles. The evidence of dust contamination is shown by deviations in the Zimm plot at angles below 45°. Tabor [43] reviewed the preparation of liquids and solutions for light scattering.

APPLICABILITY

The light scattering technique is applicable to any polymer that can be dissolved in a solvent whose neat refractive index is sufficiently different from that when the polymer is dissolved in it.

For polymers whose $M_W < 10,000$, the intensity of light scattering from the solution differs so little from the neat solvent that the determination is not precise. For polymers whose $M_W > 10,000$, the need to measure the light scattering at very small values of θ is beyond the capability of many older instruments. Special treatment is required for mixed solvent systems, copolymers, higher polymers, and polyelectrolytes.

One of the most serious situations affecting the light scattering measurement is contaminants or impurities in the solvent and/or solution. The method is inapplicable if the solvent and/or solution cannot be clarified.

ACCURACY AND PRECISION

A precision of ±5% can be obtained for repeat light scattering measurements if care is taken. The precision in M_W is approximately the same, as the scattered intensity and M_W are proportional.

Accuracy in the determination depends on the care with which the glassware is cleaned, how solvents are purified, how solutions are prepared and clarified, the calibration of the photometer, the determination of the specific refraction increment (dn/dc), etc. Consequently, the accuracy is often not better than ±10%.

SAFETY PRECAUTIONS

Safety glasses must be worn in the laboratory at all times. Safety gloves should be worn when handling solvents.

Material safety data sheets (MSDS) for all chemicals being used must be read prior to beginning the experiment. All chemicals should be considered hazardous from a standpoint of flammability and toxicity.

Organic solvents should always be used in well-ventilated areas and, when possible, in small quantities. Avoid skin or eye contact and breathing organic solvent vapors. The photomultiplier tube (PMT) of most light scattering photometers is powered by dangerously high voltages (~1 kV). The circuitry is usually well protected, but caution should be exercised. Do not attempt any instrument repairs; refer any repairs of the instrument to the laboratory instructor. Avoid exposing the phototube to high light intensity when voltage is applied to it as such exposure can result in permanent damage to the PMT.

APPARATUS

1. Light scattering photometer, with absolute calibration carried out in advance plus appropriate neutral filters, and data collection system
2. Refractometer or differential refractometer for measuring the refractive index of solutions
3. 0.1-μm Millipore filter or equivalent or ultrafine sintered glass filter
4. 100-ml stoppered volumetric flask; ultraclean
5. 20-ml Erlenmeyer flasks; stoppered
6. Cells for the photometer; ultraclean and dry
7. Aluminum weighing pans
8. Graduated or volumetric pipettes; ultraclean and dry
9. Analytical balance capable of weighing to 0.1 mg
10. A laminar flow clean air station is recommended
11. Ultrasonic cleaning bath (optional)
12. Magnetic stirrer (optional).

REAGENTS AND MATERIALS

1. Poly(styrene) prepared in experiments 1–4
2. Commercial poly(styrene) or narrow molecular weight distribution polymer with $M_W > 50,000$
3. Toluene, spectroscopic grade
4. 12-Tungstosilecic acid (12-TSA), $H_4SiW_{12}O_{40}$, $M_W = 2879$ (Note: Review Ref. 41 for special handling of this material.)
5. 2-Butanone, spectroscopic grade

PREPARATION

The pure solvents and polymer solutions should be prepared and filtered prior to the start of the experiment. It may take several hours to prepare the solutions. It is necessary to filter the solutions prior to use as dust and other extraneous particles in solution will scatter light and cause the results to be uninterpretable [16,18,41].

The preparation of the solvent and polymer solutions is given below [16,18,41].

1. All pieces of equipment to be used in the experiment should first be rinsed with fresh solvent and then with a small portion of filtered solvent.
    a. Filter toluene (1000 ml) through a fine or ultrafine sintered glass filter. The first few milliliters coming through are used to clean out the

receiving flask. The next few milliliters are used to rinse the Erlenmeyer flask, which will be used for storage of the solvent.

  b. The filtered solvent should be transferred to the storage Erlenmeyer flask, which is then stoppered with a glass stopper or an aluminum foil wrapped cork.

2. Filter and store 2-butanone or other reference solvent (~150 ml) (depending on the type of light scattering device used) as described earlier.
3. Prepare the polymer stock solution. The amount of polymer stock solution needed is determined by the size of the sample cells of the light scattering device and the refractometer (Note 1). Solution sizes may be adjusted accordingly.

  a. Take an aluminum weighing pan.
  b. Weigh 1 g of poly(styrene) into the pan and record the weight.
  c. Rinse a 250-ml Erlenmeyer flask with the filtered solvent.
  d. Add the poly(styrene) to the flask.
  e. Add approximately 100 ml of filtered solvent to the same flask to dissolve the polymer. Stopper the flask.
  f. Stir the flask periodically or magnetically until the polymer dissolves.

4. Now filter the polymer solution using Millipore filtering equipment (Note 2).

  a. Filter the polymer solution into a rinsed Erlenmeyer flask.

5. Determination of the polymer concentration.

  a. Weigh an aluminum weighing pan. Record the weight.
  b. Place a 10-ml aliquot of the filtered polymer solution into the weighed aluminum pan.
  c. Evaporate the solvent in such a way as to avoid dust or extraneous contamination. (A vacuum oven may be used.)
  d. Reweigh the aluminum pan containing the dried polymer. Record the weight.
  e. Calculate the exact concentration of the polymer.

6. Preparing the various polymer solutions.

  a. Rinse four 100-ml volumetric flasks and their stoppers with the filtered solvent.
  b. Using rinsed and dried pipettes, add 20, 15, 10, and 5 ml of the filtered polymer stock solution.
  c. Add filtered solvent to the flasks up to the dilution mark. Stopper the flasks. This gives polymer solutions of approximately 2×10^{-3}, 1.25×10^{-3}, 1×10^{-3}, and 5×10^{-4} g/ml.

7. Measurement of the refractive index.

  a. The refractive index of the solvent and the polymer solutions may be measured on a variety of instruments, such as an ordinary refractometer, a differential refractometer, or a Rayleigh interferometer. The instrument manual should be consulted for appropriate use. If the differential refractometer is used, the absolute refractive index of the solvent must be known, which can be found in standard reference works.

PROCEDURE

It is recommended that the instrument instructions and ASTM Procedure D4001-93 [41] be read prior to beginning the experiment. Allow at least 3 hr for the experiment. The solutions should be prepared and filtered prior to the start of the experiment. The photometer should be turned on at least 1 hr prior to the start of the experiment.

Calibration of Light Scattering Photometer

Calibration is required to convert measurements of scattered light intensity from arbitrary to absolute values, an essential step in the calculation of molecular weight. Fortunately, because the calibration constant of most photometers remains stable for long periods of time, the calibration procedure need be carried out only infrequently. Should it need to be calibrated, the procedure described in ASTM D4001-93 or that of the instrument vendor should be followed.

Measuring the Scattering of the Pure Solvent

a. Be sure that the photometer is prepared for measurement and stabilized.
b. Fill the cleaned scattering cell with filtered solvent; insert it in the instrument and align as required.
c. Select the wavelength-isolating filter to be used. Turn the detector to the specified angle and set the level of high voltage or adjust the slit openings to provide an appropriate solvent reading. Record these settings. In subsequent steps, do not readjust these variables, but change amplifier gain by known factors or insert neutral filters of known transmittance as required to maintain readings on scale.
d. Wait 10–15 min to allow the sample to equilibrate within the photometer.
e. Read and record the scattered intensity at angles of 30°, 90°, 150°, and at least six other angles, symmetrically placed with respect to 90°.

Measuring the Scattering of the Reference Material

a. Turn the phototube to the specified reference angle.
b. Depending on the photometer, adjust the amplifier gain or insert neutral filter as required.
c. Insert the reference standard.
d. Read and record the indicated reference intensity.

Measuring the Scattering of the Polymer Solutions

a. Using clean, dry, dust-free pipettes, add the first polymer solution to the cell.
b. Read and record the scattered intensity at the same angles used earlier.
c. Rinse the cell with filtered solvent.
d. Repeat steps a to c with the remaining polymer solutions.

FUNDAMENTAL EQUATIONS

Rayleigh Ratio

$$R_\theta = k_w (n_S/n_w)(I_S/I_r)$$

Debye Constant

$$K = 2\pi^2 n^2 (dn/dc)^2/N_0\lambda^4$$

Average Molecular Weight

$$M_W = 1/K(c\Delta R_\theta)_{c\to 0,\ \theta\to 0}$$

Second Virial Coefficient

$$A_2 = (½) K [(c/\Delta R_\theta)c_2 - (c/\Delta R_\theta)c_1] / (c_2 - c_1)$$

Radius of Gyration

$$\langle R_g^2 \rangle^{1/2} = [(3\lambda^2/16\pi^2 n^2) \cdot (\text{slope/intercept})]^{1/2}$$

CALCULATIONS

1. Tabulate values of scattered intensity, corrected for the dark current, for each scan and $\theta = 30, \ldots, 150°$. Include values for the reference. Correct all readings for differences, if any, in amplifier gain or filters used. Tabulate as in Refs 15, 16, and 41
2. Calculate ΔR_θ and $c/\Delta R_\theta$ for each angle and concentration used. Tabulate as in Refs. 15, 16, and 41
3. Compare the highest concentration used to the range of $\sin^2(\theta/2)$ covered and choose an appropriate value of k. Calculate $\sin^2(\theta/2) + kc$ for each angle and concentration used. Tabulate as in Refs. 15, 16, and 41
4. Prepare the Zimm plot and read the intercept and the slopes of the $c = 0$ and $\theta = 0$ lines.
5. Calculate the Debye constant, K.
6. Calculate $\overline{M}_W$ and the radius of gyration from the Zimm plot.

REPORT

1. Describe the experiment and apparatus, including the following information: (a) sample identification; (b) conditioning of the sample, if any; (c) solvent, temperature, and instrument used; (d) filtration technique; (e) basic data such as wavelength, dn/dc, n, vertically polarized or unpolarized light, nature of reference, and calibration constant; and (f) correction factors and any basic data (absorbance, depolarization) used in deriving them.
2. Include the Zimm plot, the table of data, and the calculated values of M_W, A_2, and $\langle R_g^2 \rangle^{1/2}$ with the appropriate units for each.
3. Answer the following questions:
   a. Is the Zimm plot rectilinear?
   b. If not, what conclusions may be drawn from the nature of the distortions?
   c. If the polymer is branched instead of linear, what changes would be required in the procedure and results?
   d. If the polymer were a copolymer or a polyelectrolyte, what change would be required in the procedure and results?

NOTES

1. The concentration of the stock solution can be estimated as follows: for a polymer of $M_W = 100{,}000$, in a solvent such that $dn/dc \approx 0.2$ ml/g, the stock solution should be in the range from 10 to 20 g/liter.
2. Two common types of filtering apparatus are used. A fine or ultrafine sintered glass may be used to filter the solvent. However, if it is used to filter the polymer solution, it may become clogged. Using Millipore

filtering equipment along with 0.1- to 1.0-μm filters usually prevents clogging with the polymer. If the filtering device uses a syringe mounted filter, care should be taken not to use excessive pressure, as this may rupture the filter.

ACKNOWLEDGMENTS

The authors thank Drs. Richard Perrinaud, Elf Atochem North America, and James S. Holten for proofreading and making corrections.

REFERENCES

1. Lord Rayleigh, *Philos. Mag.* **41**(4), 447 (1871).
2. Lord Rayleigh, *Philos. Mag.* **12**(5), 81 (1881).
3. A. Einstein, *Ann. Phys.* (*Leipzig*), **33**, 1275 (1910).
4. M. Smoluchowski, *Ann. Phys.* (*Leipzig*), **33**, 205 (1908).
5. M. Smoluchowski, *Philos. Mag.* **23**(6), 165 (1912).
6. P. Debye, "The Collected Papers of P. Debye," Wiley-Interscience, New York, 1954.
7. P. Debye, in "Light Scattering from Dilute Polymer Solutions" (D. McIntyre and F. Gornick, eds.), p. 13, Gordon and Breach, New York, 1964.
8. P. Debye, *J. Appl. Phys.* **15**, 338 (1944).
9. I. A. Katime and J.R. Quintana, "Comprehensive Polymer Science: The Synthesis, Characterization, Reactions and Applications of the Polymers" (G. Allen and J. C. Bevington, eds.), Vol. 1, Chap. 5, p. 103, Pergamon Press, New York, 1989.
10. P. Kratochvil, "Classical Light Scattering from Polymer Solutions" (A. D. Jenkins, ed.) Polymer Science Library 5, Elsevier, New York, 1987.
11. "Light Scattering from Dilute Polymer Solutions" (D. McIntyre and F. Gornick, eds.), Gordon and Breach, New York, 1964.
12. P. Debye, *J. Phys. Colloid. Chem.* **51**, 18 (1947).
13. H. E. Eisenberg, "Biological Macromolecules and Polyelectrolytes in Solution," Chap. 4, Oxford University Press, Oxford, 1976.
14. M. B. Huglin, ed., "Light Scattering from Polymer Solutions," Academic Press, London, 1972.
15. F. W. Billmeyer, Jr., "Textbook of Polymer Science," 3rd Ed., Wiley, New York, 1984.
16. E. A. Collins, J. Bareš, and F. W. Billmeyer, Jr., "Experiments in Polymer Science," Wiley-Interscience, New York, 1971.
17. P. C. Hiemenz, "Polymer Chemistry: The Basic Concepts," Dekker, New York, 1984.
18. E. M. Pearce, C. E. Wright, and B. K. Bordoloi, "Laboratory Experiments in Polymer Synthesis and Characterization," The Pennsylvania State University, University Park, PA, 1982.
19. G. V. Schulz and M. Lechner, in "Light Scattering from Polymer Solutions" (M. B. Huglin, ed.), Chap. 12, Academic Press, London, 1972.
20. H. H. Lewis, R.L Schmidt, and H.L. Clever, *J. Phys. Chem.* **74**, 4377 (1970).
21. B. M. Fechner and C. Strazielle, *Makromol. Chem.* **160**, 195 (1972).
22. T. G. Scholte, *Eur. Polym. J.* **6**, 1063 (1970).
23. T. G. Scholte, *J. Polym. Sci. A-2* **9**, 1553 (1971).
24. E. P. Pittz, J. C. Lee, B. Bablouzian, R. Townend, and S. N. Timasheff, *Methods Enzymol.* **27**(D), 209 (1973).
25. M. B. Huglin, *Top Curr. Chem.* **77**, 141 (1978).
26. A. J. Hyde, in "Developments in Polymer Characterization, Volume 1" (J. V. Dawkins, ed.), Applied Science, London, 1978.
27. B. Chu, "Laser Light Scattering," Academic Press, New York, 1974.
28. B. H. Berne and R. Pecora, "Dynamic Light Scattering," Wiley, New York, 1976.
29. K. A. Stacey, "Light Scattering in Physical Chemistry," Butterworths, London, 1956.
30. I. L. Fabelinskii, ed., "Molecular Scattering of Light," Plenum Press, New York, 1968.

31. E. F. Casassa and G. C. Berry, in "Techniques and Methods of Polymer Evaluation" (P. E. Slade, ed.), Vol. 4, Chap. 5, Dekker, New York, 1975.
32. C. Tanford, "Physical Chemistry of Macromolecules," Chap. 5, Wiley, New York, 1961.
33. M. Kerker, "The Scattering of Light and Other Electromagnetic Radiation," Academic Press, New York, 1969.
34. H. Morawetz, "Macromolecules in Solution," Chap. 5, Wiley, New York, 1965.
35. P. J. Flory, "Principles of Polymer Chemistry," Chap. 7, Cornell University Press, Ithaca, NY, 1953.
36. H. G. Elias, "Macromolecules," Vol. 1, p. 311, Plenum Press, New York, 1984.
37. G. D. Patterson and J. P. Latham, *Macromol. Rev.* **15,** 1 (1980).
38. B. H. Zimm, in "Light Scattering from Dilute Polymer Solutions" (D. McIntyre and F. Gornick, eds.), pp. 149 and 157, Gordon and Breach, New York, 1964.
39. B. H. Zimm, *J. Chem. Phys.* **16,** 1093 (1948).
40. B. H. Zimm, *J. Chem. Phys.* **16,** 1099 (1948).
41. ASTM D 4001-93, "Test Method for Determination of Weight-Average Molecular Weight of Polymers by Light Scattering," ASTM, West Conshohocken, Pennsylvania, 1997.
42. H. Utiyama, in "Light Scattering from Polymer Solutions" (M. B. Huglin, ed.), Chap. 4, Academic Press, London, 1972.
43. B. E. Tabor, in "Light Scattering from Polymer Solutions" (M. B. Huglin, ed.), Chap. 1, Academic Press, London, 1972.

EXPERIMENT 20

End Group Analysis

INTRODUCTION

The high molecular weight of a polymer is one of the most immediate consequences of the chain structures of these molecules. The ends of polymer chains sometimes consist of groups different from the monomer units that make up the body of the polymer molecule [1]. Although the analysis of these end groups is primarily useful for the determination of molecular weights [1–8], it has proved useful for studying the kinetics of polymerization and depolymerization [1]. End group analysis can also be applied to studies of polymer inhibition [1,9–12]. Work has been done using transfer agents containing elements that can easily be analyzed, thus allowing the control of molecular weight to be monitored [1,8,9,13]. Consequently, end group analysis is an important method for polymer characterization [9].

A wide range of chemical and physical techniques have been employed to the determination of functional groups as well as end groups [9]. The chemical techniques comprise methods based on halogenation, titration, saponification values, phthalation, acetylation, hydrogenation, and colorimetric procedures [9]. Physical methods comprise procedures based on infrared spectroscopy, Raman spectroscopy, ultraviolet spectroscopy, nuclear magnetic resonance spectroscopy [14], pyrolysis, or alkali fission of the polymer followed by gas chromatography [9]. The use of radioisotopes in this connection also extends the range of the molecular weights that can be determined by this approach [8,15]. This chapter is concerned with molecular weight determination based on end group analysis using titration [1–6].

THEORY

Molecular Weight Determination

The theoretical background for the termination of molecular weight of condensation or addition polymers by end group analysis can be found elsewhere [1,4]. Estimation of the total end group content by either chemical or physical methods and of the number average molecular weight, M_n, by a method such as osmometry permits the calculation of the average number of end groups per molecule, which is a measurement of the degree of branching, provided each branch terminates with the kind of end group in question [1].

Linear Polymers

For linear polymers, determination of end groups gives the number average molecular weight, M_n [1]. The technique has the advantage of not requiring calibration against another method [1]. However, successful application of end group analysis requires a knowledge of the nature of the terminal groups, the number of the groups per molecule, and methods capable of accurate estimation of the end group in question [1].

Condensation Polymers

Linear condensation polymers are most suited for end group analysis because of the certainty that all chains are terminated by reactive end groups [1,2]. Condensation polymers such as poly(esters) and poly(amides) are especially well suited to molecular weight determination by end group analyses. The chain ends in these molecules consist of unreacted functional groups, as well as having relatively low molecular weights. Using poly(amides) as an example, the following end groups are possible:

a. If a poly(amide) is prepared in the presence of a larger excess of diamine, the average chain will be capped by an amine group at each end:

$$H_2N-R(NHCO-R)_m\ NH_2.$$

In this case only the amine can be titrated and the two ends are counted per molecule.

b. A poly(amide) such as poly(caprolactam) is a linear molecule with a carboxyl group at one end and an amino group at the other:

$$HOOC-(CH_2)_5[NHCO(CH_2)_5]_n\ NH_2.$$

In this case there is one functional group of each kind per molecule. It is usually necessary to determine both end groups present in these polymers, as there are several reasons not to assume equal contents of each type of group, even when the monomers have been initially equal[1,2].

c. If the poly(amide) is prepared in the presence of a large excess of dicarboxylic acid, the average chain will have a carboxyl group at each end:

$$HOOC-R(CONH-R)_n\ COOH$$

Only acid groups are titrated and two ends are counted per molecule [1,2].

In addition to these examples, the following questions must also be considered when calculating the molecular weight from end groups.

1. Have any ring structures been formed from the reaction of two ends of the same molecule?
2. Can branching be ruled out? If a trifunctional reactant has been added, it provides more than two end groups per molecule. This makes end group analysis useful for determining the extent of branching but does not ordinarily permit molecular weight determination [1,2].
3. Have any amine groups been acetylated by the reaction with an acid catalyst?
4. Has any dicarboxylation occurred as a result of elevated temperatures?

Vinyl Polymers

Vinyl polymers are generally unsuitable for end group analysis because the free radical mechanisms of their formation may leave active end groups on one or both ends [1]. Also, the number of such ends in vinyl polymers is very small because of the much higher molecular weights (above 50,000) usually produced [1].

Other Functional Groups

Alcoholic Hydroxyl Groups

Chemical methods used for the determination of hydroxyl groups or alcoholic constituents in polymers are based on acetylation [16–18], phthalation [18], and reaction with phenyl isocyanate [18,19] or, when two adjacent hydroxy groups are present in the polymers, by reaction with potassium periodate [9,17]. Alcoholic hydroxyl groups may be found in the following polymers: (1) poly(ethylene terephthalate) (PET) [20], (2) poly(methyl acrylate), [21], (3) poly(methyl methacrylate) [21], and (4) polyhydric alcohols in hydrolysates of poly(ester) resins [22].

Phenolic Hydroxyl Groups

These hydroxyl groups in polymers are also usually determined by acetylation [9,21] or bromination [23]. However, it should be noted that acetylation with acid anhydrides and acyl chlorides that only total hydroxyl groups in these resins can be determined [7]. Aromatic sulfonyl chlorides, however, react selectively with phenolic hydroxyls [26].

Epoxy Groups

For detection and quantitative determination of small quantities of epoxy groups, these groups may be reacted with dinitroarene sulfonic acids [7,20,21,24], which react nearly as rapidly as hydrogen halides.

Olefinic Double Bonds

Anderson [25] determined the distribution of olefinic bonds in elastomers after derivitization with 2,4-dinitrobenzenesulfonyl chloride using a gel permeation chromatograph equipped with a photometer operating at 254 nm. It is also possible to determine olefinic linkages with a preliminary epoxidation, followed by a method to analyze for the epoxy [26].

The residual double bonds of poly(methyl acrylate) have been determined by bromination [9,27]. Bromination is accomplished through the addition of potassium bromide to potassium bromate in acidic medium [9]. Styrene–butadiene copolymers contain residual double bonds. The butadiene content of the copolymer has been determined by an iodine monochloride titration procedure [9].

Carbonyl Groups

One of the common chemical methods for determining carbonyl compounds consists of converting them into hydrazones [7]. This has been used for (1) oxycellulose [28], (2) nylon-6 and nylon-6,6 [29], (3) dehydrogenated poly(vinyl chloride) [30], (4) in irradiated polyethylene films [31], and (5) grafted poly(ethylene glycol [32].

Mercapto Groups

Mercapto groups in poly(caprolactam) fibers having disulfide and alkalene sulfide cross-links have been determined by swelling the sample with methanol and titrating the suspended strips with an alcoholic silver nitrate solution [1,33].

Methodology

The terminal groups of a polymer chain are different from the repeat units that characterize the rest of the molecule. If some technique of analytical chemistry can be applied to determine the number of these end groups in the polymer sample, then the average molecular weight of the polymer may be evaluated. The concept is no different than the equivalent procedure applied to low molecular weight compounds. The following steps outline the experimental and computational aspects of the procedure [2].

  a. The mass of the sample is determined using an analytical balance.
  b. The sample is made up using volumetric flasks.
  c. A suitable functional group is assayed in the same sample using an acid/base titration and a method for end point determination.
  d. From the volume and concentration of the base, the number of equivalents of the neutralized acid group is readily calculated.
  e. The number of grams in a sample divided by the number of equivalents in the same sample gives the gram equivalent weight of the material.
  f. If the number of equivalents per mole is known, the molecular weight is calculated from the equivalent weight by multiplying the latter by the number of equivalents per mole.

APPLICABILITY

End group analysis are restricted to relatively low molecular weight polymers. Chemical procedures for counting end groups are usually considered inadequate for polymers of molecular weights in excess of 20,000 to 30,000 [1]. The sensitivity of the method decreases as the molecular weight of the polymer increases (see Note 1).

ACCURACY AND PRECISION

Accuracy and precision depend on the propagation of error starting from the error in weighing, volumetrically preparing the sample, and delivering the titrant to the sample.

In general, the method is precise to about 2% of the end group concentration. The corresponding uncertainty in $\overline{M}_n$ will vary from case to case because of the reciprocal relation between the two.

SAFETY PRECAUTIONS

Safety glasses must be worn in the laboratory at all times. Appropriate safety gloves and other personal protection equipment must be used to prevent skin contact. Material safety data sheets (MSDS) must be read before handling the chemicals in these experiments. All chemicals should be considered hazardous. Certain preparations of the polymers should be performed in a well-ventilated hood.

APPARATUS

Procedure I: Amine End Groups [5,34–36]

  a. Three 100-ml three-neck flask with two stoppers each
  b. Heating mantles and controllers

c. Condenser to fit three-neck flask
d. 5-ml microburettes with 0.01-ml graduation
e. Stirring motors or magnetic stirrers
f. (Optional) Conductance bridge and conductometric cell with platinum black electrodes
g. (Optional) Cryogenic sample grinder
h. Small liquid nitrogen dewar.
i. Small laboratory mill
j. Balance capable of measuring to 0.0001 g
k. Aluminum weighing pan or weighing paper

Procedure II: Hydroxyl End Groups [5,16,37–39]

a. 250-ml iodine flask
b. 50-ml burette
c. 10-ml pipettes
d. Hot plate with magnetic stirrer
e. Balance capable of measuring to 0.0001 g
f. Aluminum weighing pan or weighing paper

REAGENTS AND MATERIALS

Procedure I: Amine End Groups [5,34–36]

a. Poly(hexamethylenesebacamide), nylon (6,10) from Experiment 9
b. Phenol, crystal, reagent grade, free flowing (see Preparation step c1)
c. Methanol, reagent grade (see Preparation step c2)
d. Hydrochloric acid, standardized, in one or more of the following concentrations: 0.5, 0.1, 0.2, 0.5, or 1.0 N (see Preparation step d)
e. Thymol blue indicator (thymosulfonphthalein), 0.1% in distilled water
f. Liquid nitrogen (optional)
g. Stop cock grease

Procedure II: Hydroxyl End Groups [5,16,37–39]

a. Poly(tetramethylene glycol) or another hydroxy-terminated polymer
b. Sodium hydroxide, reagent grade (see Preparation step e)
c. Methanol, reagent grade (see Preparation step e)
d. Potassium acid phthalate, primary standard grade (see Preparation step e)
e. Acetic anhydride, reagent grade (see Preparation step f)
f. Pyridine, reagent grade (see Preparation step f)
g. Phenolphthalein, cresol red, and thymol blue indicators (see Preparation step g)
h. *n*-Butanol, reagent grade

SAMPLE PREPARATION

Procedure I: Amine End Groups [5,34–36]

a. Procedure I takes at least 3 hr.
b. For Procedure I, the polyamide must be broken or ground into fine pieces no longer than 1 mm in dimension or dissolution is too slow for

the time allotted. The sample should be ground ahead of time. Because the poly(amide) is a tough material, it should be cooled to liquid nitrogen temperature and crushed or ground in a small laboratory mill.

c. Poly(amides) are extremely sensitive to water; care must be taken to avoid exposure. Reagents must be water free.

1. If reagent-grade phenol from a previously unopened container is not available for use in Procedure I, it will be necessary to distill the phenol from 1 g/liter BaO, collecting the constant boiling portion of the distillate.
2. If reagent-grade methanol from a previously unopened container is not available, it will be necessary to distill methanol for use in Procedure I from 1 g/liter KOH (pellets), discarding the first 10% of the distillate.

d. Standardized HCl for use in Procedure I can be purchased from laboratory supply houses or prepared by usual analytical techniques. Convenient titers are obtained by using 1 N acid for the sample polymerized 10 min, 0.5 N for 20 min, 0.2 N for 30 min, 0.1 N for 45 and 60 min, and 0.05 N for 90 min. For conductometric titration, these concentrations must be doubled.

Procedure II: Hydroxyl End Groups [5,16,37–39]

e. Prepare ahead of time 0.40 N methanolic NaOH. Dissolve 16 g of NaOH in a minimum amount of distilled water. Dilute to 1 liter with methanol and allow to stand overnight. Standardize against primary standard grade potassium acid phthalate using a phenolphthalein indicator.

f. Prepare acetylating agent ahead of time. Dissolve the amount of acetic anhydride calculated below in pyridine to a total volume of 250 ml. Calculate the volume of acetic anhydride as

$$68.85 \times N,$$

where N is the normality of the NaOH from Preparation step d. This is calculated to make a reagent giving a blank titration of slightly less than 50 ml standard NaOH.

g. Prepare the mixed indicator by adding 1 part 0.1% aqueous cresol red to 3 parts 0.1% aqueous thymol blue; both indicators having been neutralized with NaOH.

PROCEDURE

Procedure I: Amine End Groups [5,34–36]

a. Place 35 g of phenol and 15 g (19 ml) methanol in each of three 100-ml three-neck flasks.
b. Tare an aluminum weighing pan.
c. Accurately weigh three samples (1.5–2 g) of finely ground poly(amide). Record the weight.
d. Add the poly(amide) to each flask.
e. Fit each flask with two stoppers and condenser.
f. Heat with refluxing until the sample is dissolved completely.
g. Cool the flasks to room temperature.
h. Replace condenser with stirrer.

i. Add 0.2 ml of thymol blue indicator solution to each flask.
j. Add 0.2-ml increments of standardized HCl solution. Record the concentration of HCl used.
k. Titrate to a pink end point. Record the amount of HCl added.
l. (Optional) For conductometric titration, do not add thymol blue indicator solution.

1. Add 5 ml of distilled water and mix.
2. Titrate using 0.05-ml increments of standardized HCl.
3. The end point of the titration is when the curve of conductivity vs ml HCl changes slope.

Procedure II: Hydroxyl End Groups [5,16,37–39]

Hydroxyl Equivalent

1. Weigh and tare a 250-ml iodine flask.
2. Accurately weigh the amount recommended below of the polymer to be analyzed into the 250-ml iodine flask. Record the weight of the polymer.

M_n	Weigh in (g)
400	1.2
800	1.8
1200	2.0
1600	2.3
2000	3.0

3. Repeat steps 1 and 2 for another 250-ml iodine flask.
4. Prepare two empty flasks as blanks. (*Note:* Follow steps 5–14 for each of the flasks prepared in steps 1–4.)
5. Accurately pipette 10.0 ml acetylating reagent (Preparation step f) into the flask and stopper immediately.
6. Place a magnetic stirring bar into each of the flasks.
7. Place the flasks on stirring hot plates.
8. Add 10 ml distilled water and 10 ml pyridine to each of the flasks.
9. Heat the flasks for 5 min at approximately 100°C.
10. Remove the flasks from the stirring hot plates and cool to room temperature.
11. Add 10 ml *n*-butanol to each of the cooled flasks.
12. Add 6 drops of mixed indicator (Preparation step g).
13. Titrate with 0.04 *N* NaOH (Preparation step f) to a neutral end point.
14. Record the titrate values.

Acid Equivalent

1. Weigh and tare a 250-ml iodine flask.
2. Accurately weigh 2–3 g of polymer into the iodine flask.
3. Record the weight of the polymer.
4. Repeat steps 1–3 for another 250-ml iodine flask.
5. Prepare two empty flasks as blanks. (*Note:* Follow steps 6–11 for each of the four flasks prepared in steps 1–4.)
6. Add 25 ml pyridine and a magnetic stirring bar to each of the flasks.
7. Heat each of the flasks on a hot plate at 105–110°C until the sample is dissolved.
8. Add 10 ml of distilled water and heat for 3 min.
9. Cool the flasks to room temperature.

10. Add 10 ml *n*-butanol and 6 drops of mixed indicator (Preparation step g) to each of the flasks.
11. Titrate with 0.4 *N* NaOH (Preparation step f) to the first blue end point. Record the amount of titrant added. (*Note:* This end point fades rapidly.)

FUNDAMENTAL EQUATIONS

Procedure I: Amine End Groups [5,34–36]

$$\text{Molecular weight} = \overline{M}_n = (\text{sample wt} \times 1000)/(\text{titer, ml} \times \text{normality})$$

$$\text{Degree of polymerization} = \overline{\chi} = \overline{M}_n/M_0,$$

where M_0 is the molecular weight of the repeat unit (monomer less water).

Procedure II: Hydroxyl End Groups [5,16,37–39]

$$\text{Hydroxyl equivalent} = [(\text{titer of blank} - \text{titer of sample}) \times \text{normality}]/\text{sample wt},$$

where the sample and the blank are those of Procedure II, hydroxyl equivalent.

$$\text{Acid equivalent} = [(\text{titer of sample} - \text{titer of blank}) \times \text{normality}]/\text{sample wt},$$

where the sample and the blank are those of Procedure II, acid equivalent. Average molecular weight of PTMEG:

$$M_n = 2000/(\text{hydroxyl number} + 2 \times \text{acid equivalent})$$

CALCULATIONS

Procedure I: Amine End Groups [5,34–36]

a. Calculate $\overline{M}_n$ and $\overline{\chi}_n$ for each sample.

Procedure II: Hydroxyl End Groups [5,16,37–39]

a. Calculate the hydroxyl equivalent, acid equivalent, and M_n from average titers for duplicate samples and blanks.

REPORT

Procedure I: Amine End Groups [5,34–36]

a. Describe the experiment and apparatus in your own words.
b. Tabulate, for each sample, values of sample weight, normality of the acid used, titer, $\overline{M}_n$, and $\overline{\chi}_n$.
c. Why is it necessary to use "dry" reagents in this experiment?
d. Why is it necessary to avoid large titers in this experiment?
e. One of the difficulties encountered in performing end group analysis in the nylon system is the lack of a proper solvent. Formic acid has been shown to be a good solvent for many nylons. Would this be a good solvent in this experiment? Why or why not?
f. (Optional) Explain the change in slope of the conductivity titer curve at the end point in the conductometric titration.

Procedure II: Hydroxyl End Groups [5,16,37–39]

a. Describe the apparatus and experiment in your own words.
b. Tabulate the individual titers and their average values for each step as well as the hydroxyl and acid equivalents and the average molecular weight.

NOTE

A caution must be issued. Chemical methods for molecular weight determination become insufficiently sensitive when the molecular weight is large [6]. Spurious sources of end groups not taken into account in the assumed reaction mechanism become consequential as the molecular weight increases and the number of end groups eventually diminishes to the point where quantitative determination is impractical [6]. Consequently, the determination of molecular weights by chemical methods finds widespread use only for condensation polymers, which seldom have an average molecular weight exceeding 25,000.

ACKNOWLEDGMENT

The authors thank Dr. Herminder Sidhu, Elf Atochem North America, for proofreading and making suggestions.

REFERENCES

1. "Characterization of Polymers: Encyclopedia Reprints" (N. M. Bikales, ed.), Wiley-Interscience, New York, 1971.
2. P. C. Hiemenz, "Polymer Chemistry: The Basic Concepts," Dekker, New York, 1984.
3. E. M. Pearce, C. E. Wright, and B. K. Bordoloi, "Laboratory Experiments in Polymer Synthesis and Characterization," The Pennsylvania State University, University Park, PA, 1982.
4. M. Hellman and L. A. Wall, "Analytical Chemistry of Polymers" (G. M. Kline, ed.), Part III, Chapter V, Wiley, New York, 1962.
5. E. A. Collins, J. Bareš, and F. W. Billmeyer, Jr., "Experiments in Polymer Science," Wiley, New York, 1973.
6. P. J. Flory, "Principles of Polymers Chemistry," Cornell University Press, Ithaca, NY, 1953.
7. J. Urbański, "Applied Polymer Analysis and Characterization: Recent Developments in Techniques, Instrumentation, Problem Solving" (J. Mitchell, Jr., ed.), Chap. II-A, Hanser Publishers, New York, 1987.
8. G. F. Price, "Techniques of Polymer Characterization" (P. W. Allen, ed.), Butterworth, London, 1959.
9. T. R. Crompton, "Analysis of Polymers: An Introduction," Pergamon Press, New York, 1989.
10. C. C. Price and D. A. Durham, *J. Am. Chem. Soc.* **64**, 2508 (1942).
11. C. C. Price, *J. Am. Chem. Soc.* **65**, 2380 (1943).
12. E. L. Stanley, "Analytical Chemistry of Polymers: Analysis of Monomers and Polymeric Materials" (G. M. Kline, ed.), Interscience, New York, 1962.
13. R. M. Joyce, W. E. Hanford, and J. Harmon, *J. Am. Chem. Soc.* **70**, 2529 (1948).
14. J. C. Randall, "Polymer Characterization by ESR and NMR" (A. E. Woodward and F. A. Bovey, eds.), ACS Symposium Series 142, American Chemical Society, 1980.
15. G. B. Gerber and J. Remy-Defraigne, *Anal. Biochem.* **11**(2), 386 (1965).

16. ASTM Method D 4274-94, "Standard Test Methods for Testing Polyurethane Raw Materials: Determination of Hydroxyl Numbers of Polyols," ASTM, Philadelphia, PA, 1994.
17. J. Majewski, *Polimery* **18**(3), 142 (1973).
18. R. S. Stetzler and C. F. Smullin, *Anal. Chem.* **34,** 194 (1962).
19. L. Yu Spirin and T. A. Yatsimirskaya, *Vysokamal. Soedin. Ser* **A15**(11), 2595 (1973).
20. H. Zimmerman and C. Kolbig, *Faserforsch. Texteltechn.* **18,** 536 (1967).
21. J. Urbański, W. Czerwiński, K. Janicka, F. Majewska, and H. Zowall, "Handbook of Analysis of Synthetic Polymers and Plastics," Ellis Horwood, Chichester, England, 1977.
22. J. Gasparić and J. Borecký, *J. Chromatogr.* **5,** 466, (1961).
23. H. Sidhu, private communication.
24. H. Dannenburg and W. R. Harp, Jr., *Anal. Chem.* **28,** 86 (1956).
25. J. N. Anderson, *J. Appl. Polym. Sci.* **18**(9), 2819 (1974).
26. P. Dreyfuss and J. P. Kennedy, *Anal. Chem.* **47,** 771 (1975).
27. E. C. Kuryanikov, R. V. Vizgert, A. A. Berlin, and L'viv Vijn, *Politekh. Inst.* **57,** 36 (1971).
28. H. S. Blair and R. Cromie, *J. Appl. Chem. Biotechnol.* **27**(4), 205 (1977).
29. V. Rossbach, *Schriftenr. Dtsch. Wollforschungsinst.* (Tech. Hochsch. Aachen), **80,** 134 (1979).
30. T. Morikawa, *Kubonshi Kagaku* **24,** 592 (1967).
31. Z.-H. Chen, Y.-C. Song, B.-Y. Han, C.-D. Jiang, X.-H. Zhang, and F.-J. Liu, *Kao Fen Tzu Tung Hsun* **1980**(6), 378; *Chem. Abstr.* **95,** 81849 (1981).
32. J. J. Villenave, R. Jaouhari, C. Filliatre, and M. Barathart, *Analysis* **11**(3), 136 (1983).
33. S. D. Bruck and S. M. Bailey, *J. Res. Natl. Bur. Std.* **A66,** 185 (1982).
34. ASTM D 2073-92, "Standard Test Methods for Total, Primary, Secondary, and Tertiary Amine Values of Fatty Amines, Amidoamines and Diamines by Referee Potentiometric Method," ASTM, Philadelphia, PA, 1992.
35. ASTM 2074-92, "Standard Test Method for Total, Primary, Secondary, and Tertiary Amine Values of Fatty Amines by Alternative Indicator Method," ASTM, Philadelphia, PA, 1992.
36. ASTM 2076-92, "Standard Test Methods for Acid Value and Amine Value of Fatty Ammonium Chlorides," ASTM, Philadelphia, PA, 1992.
37. ASTM E 335-96, "Standard Test Method for Hydroxyl Groups by Pyromellitic Dianhydride Esterification," ASTM, Philadelphia, PA, 1996.
38. ASTM E 326-96, Standard Test Method for Hydroxyl Groups by Phthalic Anhydride Esterification," ASTM, Philadelphia, PA, 1996.
39. ASTM E 222-94, Standard Test Methods for Hydroxyl Groups Using Acetic Anhydride Acetylation," ASTM, Philadelphia, PA, 1994.

EXPERIMENT 21

X-Ray Diffraction

INTRODUCTION

The diffraction of X-rays has become a powerful tool in the study of the structure of polymers [1–16]. The wavelengths of X-rays are comparable to the interatomic distance in crystals (0.5–2.5 Å). Two primary diffraction methods are used to study polymers: (a) wide angle X-ray scattering (WAXS) and (b) small angle X-ray scattering (SAXS). WAXS has been used for decades to study the structural properties of polymers [1–5,9,12–15]. The value of angles used in WAXS is from 5° to 120°. The primary information generally obtained from a diffraction experiment is the structure of semicrystalline polymers, with a range of interatomic distances of 1 to 50 Å. As the crystallinity of polymers is often low, the width of the diffraction peaks in WAXS gives information on the size of the crystals. From the measurement of relative intensities of diffraction peaks in the crystalline part and the diffusion halo from the amorphous part, the crystalline content of the polymer may be deduced (Fig. 21.1). WAXS has also been used to provide information on the number of repeat units per turn in helical structures that are typical of linear polymers, the length of the repeat unit along the fiber axis, and the degree of orientation [3].

The value of angles used in SAXS is from 1° to 5°. SAXS provides information on greater interatomic distances: from 50 to 700 Å. Consequently, SAXS is useful in detecting larger *periodicities* in a structure. For example, many polymeric materials crystallize with individual chains folding back and forth within a given crystalline region or crystallite [3]. Other examples of the utility of SAXS are in the study of lamellae crystallites or in the distribution of particles or voids in the material.

PRINCIPLE

Theory

X-ray diffraction (XRD) and scattering experiments involve placing the sample in the path of a monochromatized X-ray beam of low divergence. The scattered X-rays from the regularly placed atoms interfere with each other, giving strong diffraction signals in particular directions. The directions of the diffracted beams are related to the slope and dimensions of the unit cell of the crystalline lattice, and the diffraction intensity depends on the disposition of the atoms within the unit cell [5].

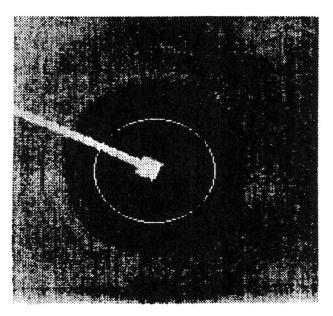

Fig. 21.1 Diffraction pattern of an anisotropic sample.

Figure 21.2 shows how the scatter pattern formed can be shown by X-rays scattered by two electrons. One electron is at the origin and the other is specified by the position vector, **r**. If s_0 and **s** are unit vectors representing the incident and scattered X-rays, respectively [5], the path difference at a removed point P is given by the dot product, $(s-s_0) \cdot r$. The phase difference between the incident and scattered X-rays is given by $(2\pi/\lambda)(s-s_0) \cdot r$, where λ is the wavelength of the X-rays [5]. A scattering vector may be defined as $S = (s-s_0)/\lambda$. The phase difference may be written as $2\pi S \cdot r$. The dimension of $|S|$ is the reciprocal length, with **S** called the reciprocal vector [5]. If s_0 is held fixed and **s** rotates, then **S** varies in both magnitude and direction. The end of the reciprocal vector, **S**, passes through a region called reciprocal space, which is the space where the diffraction pattern is found [5]. In Fig. 21.3, the circle (or sphere in three dimensions) represents the conditions for diffraction. The sphere is known as the Ewald sphere and has a radius of $1/\lambda$ (Fig. 21.3). The Ewald sphere is a useful tool in the interpretation of X-ray diffraction patterns. If the diffraction angle is chosen to be 2θ, then [5]

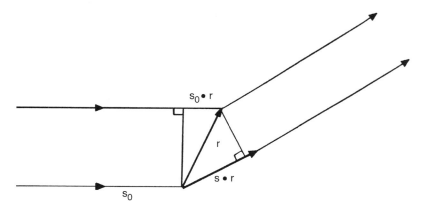

Fig. 21.2 Phase difference between waves scattered by two electrons.

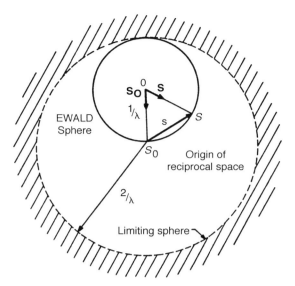

Fig. 21.3 The Ewald Sphere construction.

$$|\mathbf{S}| = 2\sin\theta/\lambda. \tag{1}$$

If **S** is replaced with $1/d$, where d represents the interplanar spacing of regularly arranged atoms, Bragg's law is obtained [5]

$$2d\sin\theta = n\lambda, \tag{2}$$

with n indicating the order of diffraction and integer values.

Amorphous Samples [3,17]

Noncrystalline or amorphous materials produce patterns with only a few diffuse maxima, which may be either broad rings or arcs if the amorphous regions are partially oriented [3]. Synthetic polymers, which are branched or cross-linked, are usually amorphous, as are linear polymers with bulky side groups, which are not spaced in a stereoregular manner along the backbone [3].

Amorphous materials can be oriented by stretching. In some instances, this orientation is also accompanied by crystallization [3]. Isoprene rubber exhibits such behavior. The positions of the maxima in the amorphous scattering pattern provide a measure of the average intermolecular spacing. Bragg's law may be used to calculate the size of the interplanar spacing [3].

Degree of Orientation [3,18]

Orientation of a polymer may be achieved by the mechanical deformation of the polymer by drawing, stretching, or rolling, which results in increased alignment of the polymer chains in the direction of the deforming force [3]. The assembly of polymer molecules is described as having a *preferred orientation*. An example of preferred orientation is found through the continuous web process such as that used to form paper or a polymer film. In both cases the tension along the film or *machine* direction is sufficient to induce partial orientation of the polymer chains.

In order to compare orientation in related samples, Hermans suggested the following equation to calculate a *degree of orientation* [3,19]:

$$F = 1/2\,(3\,\overline{\cos^2\phi} - 1), \tag{3}$$

where $\overline{\cos^2\phi}$ is the mean square cosine, averaged over all molecules, of the angle between a given crystal axis, often along the chain axis, and a reference direction, such as the fiber axis. Values of $F = 1, 0, -1/2$ describe systems with perfect, random, or perpendicular alignment of the polymer chains relative to the reference directions, respectively [3]. There are three ways to obtain an estimate of $\overline{\cos^2\phi}$ for the orientation of the chain axis to the fiber axis. One method requires measurement of the arc width of maxima on the vertical (meridional) axis of the X-ray pattern (Fig. 21.4) [3]. The vertical axis is parallel to the stretch direction of the samples. Another more accurate measurement is obtained from a radial intensity trace through the arc [3]. Third, a similar measurement of an equatorial reflection provides a measure of ϕ [3].

Crystalline Samples [3,20]

When a crystalline material is placed in a monochromatic X-ray beam, the resulting diffraction or scattering pattern depends on the sample shape or form. Simplistically, if the sample is a powder, then the scattering pattern appears as sharp concentric circles. If the sample is a fiber, then the scattering pattern appears as sharp arcs falling on "layered lines." Each maximum arises from the constructive interference between the X-rays scattered by a set of parallel planes within the crystal [3]. Rewriting Bragg's law yields

$$d_{hkl} = (\lambda/2)\sin\theta_{hkl}, \tag{4}$$

where d_{hkl} is the spacing of the lines, λ is the wavelength of the incident beam, and θ_{hkl} is the scattering angle. It is apparent from Fig. 21.5 that the angle between the emergent and scattered beams is actually 2θ. A calibrating material of known "d" spacing is often dusted on the sample to aid in the measurement of the sample's "d" spacing [3]. This superimposes a calibration ring onto the diffraction pattern of the sample. The film to sample distance, r, can be calculated as the scattering angle, θ, can be calculated from Bragg's law

$$\frac{x}{r} = \tan(2\theta), \tag{5}$$

where x is the measured radius of the calibration ring on the pattern [3].

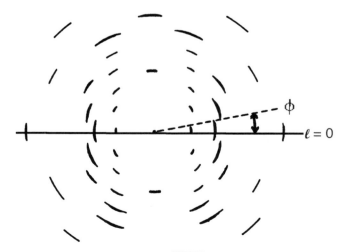

Fig. 21.4 Estimation of $\overline{\cos^2\phi}$ from a fiber pattern.

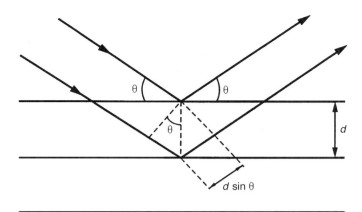

Fig. 21.5 Conditions for constructive of X-rays with a set of parallel planes of crystal.

For a fiber pattern, lines are marked through the center of the pattern parallel and perpendicular to the machine direction [3]. The line parallel to the machine direction is the meridian and the other line is the equator. The spots fall on layer lines that are roughly parallel to the equator. The equator is designated the $\ell = 0$ layer line whereas the layer line on the other side is $\ell = 1$ [3].

Experimental Methods

X-Ray Generation and Filters

X-rays may be produced by bombarding a metal target with a beam of high-voltage electrons [1,4]. This may be done inside a vacuum tube or with a synchrotron [1,4]. Choice of the target metal and the applied voltage determines the wavelength(s) of X-rays produced [1,4]. Experiments in which nearly monochromatic X-rays are produced and used are of the greatest interest; consequently, appropriate filter materials are used. Table 21.1 lists a series of common filter material used for various characteristic tube radiations.

X-Ray Detection

The original and still useful method of X-ray detection is the exposure of photographic film. In the early days of X-ray diffraction, films were also used for the quantitative recording of X-ray intensities. Densitometer measurements of optical density, D, were made of the film. The optical density is defined by

$$D = -\log_{10}(i - i_0), \qquad (6)$$

TABLE 21.1
Wavelengths (Å) and Filter Materials for the Most Common X-Ray Tubes

Target	Mo	Cu	Co	Fe	Cr
$K\alpha_2$	0.714	1.544	1.793	1.940	2.294
$K\alpha_1$	0.709	1.541	1.789	1.936	2.290
$K\beta$	0.632	1.392	1.621	1.757	2.085
Filter	Zr	Ni	Fe	Mn	V
λ_K	0.688	1.488	1.743	1.896	2.269

where i_0 and i are the intensity of the incident and transmitted light beam, respectively [4]. However, with the advent of reliable electronic counters, the use of film, though still very common in qualitative applications, was practically abandoned for qualitative X-ray intensity measurements. This was due to the ease of processing counter data, which eliminates the use of a dark room and the greater dynamic range of counters. The function of the counters is to convert the individual X-ray photons into voltage pulses. These pulses are either subsequently (a) counted or (b) integrated by the counting equipment yielding various forms of visual indication of X-ray intensity. The following types of counters are used in conventional X-ray measurements: gas-filled, scintillation, and solid state [4].

APPLICABILITY

X-ray diffraction experiments are applicable to all crystalline and amorphous solids and to all liquids and dispersions. The intensity of the X-ray scattering depends on the number of electrons in the atoms. Consequently, the obtainable information is limited in materials containing only light atoms. No information is obtained about the locations of hydrogen atoms.

ACCURACY AND PRECISION

The accuracy of a structure deduced from such spacings depends on the number of reflections observed. For a semicrystalline polymer showing only three or four reflections, only a rough estimate of the structure can be obtained. However, the structure of a metal or inorganic salt, which may show more than a thousand reflections, can be determined with very high reliability.

The precision of determining interplanar spacings is dependent on the sophistication of the equipment used. Time-consuming crystallographic techniques can give precision as good as ±0.01%.

SAFETY PRECAUTIONS

Safety glasses must be worn in the laboratory at all times. Other personal protection equipment should be worn as needed, especially appropriate safety gloves. Sodium fluoride is a deadly poison if ingested; full chemical precautions should be observed. Material safety data sheets (MSDS) should be reviewed before beginning the experiment.

Extra precaution must be taken during this experiment. Two potential hazards of X-ray diffraction experiments are the high voltage employed in producing the X-rays and the X-ray radiation itself. Students should not attempt to open the X-ray generator housing as high voltage may persist even after the instrument is turned off. The X-rays used in diffraction experiments are characterized as "soft radiation." This means that they tend to be absorbed by living tissue. The radiation can cause severe burns and blisters. Prolonged exposure may lead to the formation of tumors in exposed regions. Students should not attempt to manipulate either the sample or the film when the X-ray port is open. Students should not open a port that is blocked by a camera or lead shield. The equipment must be checked periodically for stray radiation and overall level following prescribed safety regulations.

APPARATUS

1. X-ray diffractometer with copper target, beryllium window, nickel filter, and data acquisition and processing system
2. For wide angle scattering, flat film camera with appropriate film back and intensifying screen (Fig. 21.6)
3. For small angle scattering, Statton or equivalent camera
4. For drawing fiber samples, oil bath regulated at 160°C
5. Photographic developing facilities for film used with Statton camera
6. Double-edged tape for mounting samples
7. Artist's brush for dusting NaF on sample
8. A light box for viewing
9. Ruler or appropriate tool for measuring the diffraction patterns
10. Thin-walled melting point tube (diameter 0.7 mm or less) (optional)
11. Paper clips
12. Vacuum pump
13. Tracing paper or computer scanner
14. Tweezers

REAGENTS AND MATERIALS

1. Sodium fluoride, reagent grade (calibration standard)
2. Powdered amorphous polymer such as atactic poly(styrene) or poly(methyl methacrylate)
3. Powdered linear poly(ethylene) such as Marlex 50 or equivalent
4. Poly(ethylene) plastic strap from a six pack of beverage cans
5. Poly(ethylene terephthalate) film, Mylar
6. Photographic film appropriate for wide angle scattering accessory (optional, depending on the age of the instrument, computer, and software)

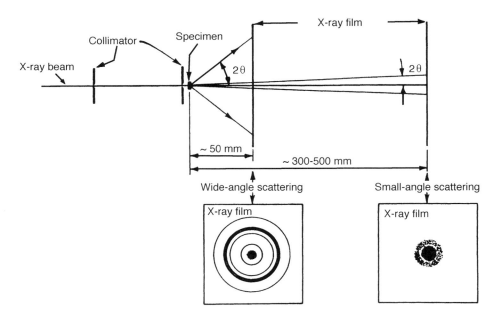

Fig. 21.6 Representative wide angle camera and specimen holder arrangement.

7. X-ray film appropriate for small angle scattering accessory (optional, depending on the age of the instrument, computer and software)
8. Paraffin wax (optional)

PREPARATION

Sample Mounting

a. Pack powder sample into a hole 1–2 mm in diameter in a piece of metal or plastic about 1.5 mm thick. Mount this piece on the specimen holder with pressure-sensitive tape so that the X-ray beam passes through (alternatively, use a thin-walled melting-point tube).
b. Mount pieces of polymer and fibers, if stiff enough to stand alone, on the specimen holder with pressure-sensitive tape. The samples should be at least 1.5 mm thick.
c. Mount several fibers on a suitable frame, such as a paper clip. Place this on the specimen holder. Be sure that the fibers are placed parallel to each other. The bundle of fibers should be at least 1.5 mm thick.

Fiber Preparation

a. Place a small amount of linear poly(ethylene) in a test tube.
b. Melt by immersing in an oil bath at 160°C for 30 min.
c. Draw fibers with tweezers.

PROCEDURE

The instrument instructions or vendor's manual should be read prior to start-up. References 1, 3, and 21–23 provide further insight into X-ray diffraction experiments. Newer equipment may be far more automated than what is described in the following experiment.

Start-up

a. Check to be sure that the ports on the X-ray tube are closed.
b. Be sure that coolant values are open.
c. Be sure that the generator is plugged in.
d. Energize the voltage circuits.
e. Insert a fresh film cassette.
f. Carefully open the X-ray port of the camera and use a piece of fluorescent material mounted on a wooden rod to check for a beam. The beam should appear as a circular spot 1–2 mm in diameter when viewed at 5 cm from the inner collimator.
g. Also check that the beam strikes the lead beam star mounted in the center of the film cassette. Close the port. The camera is now ready for use.

Wide Angle Experiments

It is suggested that the experiments be run in the following order with the following recommended exposure times:

NaF calibration	30 min
Amorphous polymer	30 min
Bulk poly(ethylene)	30 min
Poly(ethylene) fibers	60–90 min

If the small angle experiment is run simultaneously, proceed to step 1 of Small Angle Experiments.

1. Set the desired exposure time on the timer.
2. Place the specimen holder with the sample in place.
3. Load the film into the camera. Set it for exposure.
4. Open the shutter of the X-ray tube.
5. Turn the X-rays on and expose for the predetermined time.
6. Close the shutter of the X-ray tube.
7. Remove and develop the film as described by the manufacturer.
8. Repeat steps 1 to 7 for the remaining samples.
9. If the experiment is finished, turn off the high voltage.

Small Angle Experiments

If the wide angle and small angle experiments are run simultaneously, the timer is not used. For the wide angle experiment, follow steps 2 to 4 and 6 to 9, timing the exposure by clock.

1. Follow the start-up procedure.
2. Load the Statton or equivalent camera as follows:
   a. Open the camera.
   b. Insert the specimen holder.
   c. Load the film cassette in a photographic darkroom.
   d. Insert the film cassette in the camera.
   e. Replace the cover on the camera.
   f. Turn on the vacuum pump of the camera.
3. Open the shutter on the X-ray tube.
4. Start the X-ray exposure.
5. At the end of the exposure period, turn off the X-ray beam.
6. Stop the camera vacuum pump.
7. Bleed air into the camera.
8. Open the camera and remove the film cassette for development, according to manufacturer's instructions.
9. After all the experiments are completed, turn the apparatus off.

FUNDAMENTAL EQUATIONS

Bragg Equation

$$n\lambda = 2d \sin\theta,$$

where $n = 1$ for this experiment.

$$\text{Log } d_{hk0} = \log a - 1/2 \log [h^2 + k^2 (a/b)^2] \qquad (7)$$

where h and k are the first two of the Miller indices $h\ k\ l$.

CALCULATIONS

1. Calculate the exact sample film distance for the wide angle camera from the calibration sample, NaF (Note 1).

2. Calculate the average nearest neighbor distance for the amorphous polymer sample.
3. Calculate the values of d for all observed equatorial reflections for the bulk polymer and the stretched polymer sample (Notes 2 and 3).
4. Using the Hull–Davey chart (Fig. 21.7) determine the Miller indices of the reflections, the ratio b/a, and the corresponding ratio a.
5. Using Eq. (7) and the values of a/b or b/a and of log d calculated in step 4, determine a and b (Note 4).

REPORT

1. Describe the apparatus and experiment used.
2. Include the photographic prints and the small angle scattering patterns, if obtained.
3. Describe the appearance of each pattern. Are there rings or spots and are they sharp or diffuse?
4. Can the calibration ring be identified? How does the appearance of the calibration ring differ from that of diffraction maxima of the polymer?
5. If the plastic strap was used, what morphological changes arise from stretching? Do the X-ray observations agree with the mechanical changes observed?
6. Give the sample film distance obtained from the calibration experiment and the nearest-neighbor distance for the amorphous polymer.
7. Tabulate Miller indices, 2θ, and for all identified equatorial reflections.
8. Give the values of a and b and discuss them in terms of literature values.

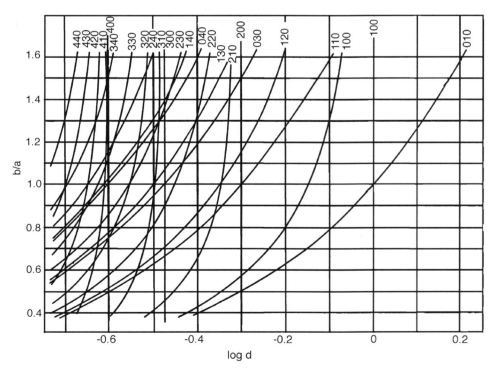

Fig. 21.7 Hull–Davey chart for determining the unit cell dimensions of an orthorhombic lattice.

NOTES

1. Note that $y/x = \tan 2\theta$ (Fig. 21.6). Search for the (200) reflections of NaF, for which $d = 2.319$ Å, assuming $x = 50$ mm. Note that the diameter of the (200) ring is $2y$. Take $\lambda = 1.541$ Å, using Bragg's law find $\sin 2\theta$ and from that obtain $\tan 2\theta$.
2. If tables of interplanar distances are available as a standard accessory to the instrument, only 2θ needs be calculated, then look up d in the tables.
3. For polymers, a problem in identifying the reflections causes a problem in determining atomic arrangements from X-ray diffraction results. Polymers do not usually exhibit reflections greater than 4 or 5. Using oriented materials greatly simplifies data.
4. To use Fig. 21.7, one plots the values of $\log d_{hkl}$ on the same scale, although not necessarily on the same range as the Hull–Davey chart using a sheet of tracing paper. Place this paper parallel to each axis until all the observed $\log d$ values fall on the chart [3]. The indices of the lines are the $h, k,$ and l indices of the corresponding spots [3]. The log d value corresponding to the 1,0 line is $\log a$, whereas $\log b$ is 0,1 [3].

REFERENCES

1. F. W. Billmeyer, Jr., "Textbook of Polymer Science," 3rd Ed., Wiley, New York, 1984.
2. G. A. Collins, J. Bareš, and F. W. Billmeyer, "Experiments in Polymer Science," Wiley, New York, 1973.
3. E. M. Pearce, C. E. Wright, and B. K. Bordoloi, "Laboratory Experiments in Polymer Synthesis and Characterization," The Pennsylvania State University, University Park, PA, 1982.
4. F. J. Baltá-Calleja and C. G. Vonk, "X-Ray Scattering of Synthetic Polymers" (A. D. Jenkins, ed.), Polymer Science Library 8, Elsevier, New York, 1989.
5. E. D. T. Atkins, "Comprehensive Polymer Science: The Synthesis, Characterization, Reactions and Applications of Polymers," (G. Allen and J. C. Bevington, eds.), Vol. 1, p. 613, Pergamon Press, New York, 1989.
6. D. P. Shoemaker, C. W. Garland, J. I. Steinfeld, and J. W. Nibler, "Experiments in Physical Chemistry," 4th Ed., McGraw-Hill, New York, 1981.
7. A. Guinier, "X-Ray Diffraction: In Crystals and Amorphous Bodies," Freeman, San Francisco, 1964.
8. H. H. Willard, L. L. Merritt, Jr., and J. A. Dean, "Instrumental Methods of Analysis," 5th Ed., Van Hostrand, New York, 1974.
9. M. L. Miller, "The Structure of Polymers," Reinhold, New York, 1966.
10. C. Tanford, "Physical Chemistry of Macromolecules," Wiley, New York, 1961.
11. International Tables for X-Ray Crystallography, Volume II, Int. Union of Crystallography, The Kynoch Press, Birmingham, 1962.
12. R. Hosemann and S. N. Bagchi, "Direct Analysis of Diffraction Matter," North Holland, New York, 1962.
13. B. K. Vainshtein, "Diffraction of X-Rays by Chain Molecules," Elsevier, Amsterdam, 1966.
14. M. Kakudo and N. Kasai, "X-Ray Diffraction of Polymers," Elsevier, Amsterdam, 1972.
15. L. E. Alexander, "X-Ray Diffraction Methods in Polymer Science," Wiley-Interscience, New York, 1969.
16. J. Formica, "Handbook of Instrumental Techniques for Analytical Chemistry" (F. Settle, ed.), Chap. 18, Prentice Hall, New Jersey, 1997.
17. H. P. Klug and L. E. Alexander, "X-Ray Diffraction Procedures," Chap. 12, Interscience, New York, 1974.
18. L. E. Alexander, *Proc. R. Soc.* **180A,** 241 (1942).

19. P. H. Hermans, "Physical Chemistry of Cellulose Fibers," p. 255, Elsevier, New York.
20. H. P. Klug and L. E. Alexander, "Physical Chemistry of Cellulose Fibers," Chap. 10, Elsevier, New York.
21. ASTM D 934-80 (reapproved 1994), "Standard Practices for Identification of Crystalline Compounds in Water-Formed Deposits by X-Ray Diffraction," ASTM, Philadelphia, PA, 1994.
22. ASTM E 1426-94, "Standard Test Method for Determining Effective Elastic Parameter for X-Ray Diffraction Measurements of Residual Stress," ASTM, Philadelphia, PA, 1994.
23. ASTM D 5380-93, "Standard Test Method for Identification of Crystalline Pigments and Extenders in Paint by X-ray Diffraction Analysis," ASTM, Philadelphia, PA, 1993.

EXPERIMENT 22

Optical Microscopy

INTRODUCTION

Microscopy is the study of fine structure and morphology of polymers with the use of a microscope [1–4]. In polymer science the term morphology refers to form and organization on a size scale above the atomic arrangement, but smaller than the size and shape of the whole sample.

It is known that the structures present in a polymer reflect the processing variables and that they greatly influence the physical and mechanical properties. Thus, the properties of polymeric materials are influenced by their chemical composition, process history, and the resulting morphology. Morphological study usually requires two preparatory steps prior to the study itself: selection of instrumental techniques and development of specimen preparation techniques. Structural observations must be correlated with the properties of the material in order to develop an understanding and applications of the material. Figure 22.1 illustrates the types of optical microscope (OM) techniques commonly used to examine polymer specimens [2].

BACKGROUND

Polymer Morphology: Single Crystals

Polymers are considered either amorphous or crystalline [5–10], although they may not be completely one or the other. It is possible to grow single crystals of polymers. The units of organization in polymer crystals are lamellae or crystals and spherulites [1,2,11–13]. Lamellae are thin, flat platelets often 100 Å thick and many micrometers in lateral dimensions [13]. A typical lamellar crystal is shown in Fig. 22.2. The size, shape, and regularity of the crystals depend on their growth conditions. The types of solvent, temperature, stress, contaminents, and growth rate are important factors [13]. The thickness of the lamellae depends on crystallization temperature and any subsequent annealing treatment [14].

Electron diffraction measurements indicate that polymer chains are generally oriented normal or very nearly normal to the plane of the lamellae [13,14]. As the molecules in the polymer are at least 1000 Å long and the lamellae are only about 100 Å thick, the most plausible explanation is that the chains are folded [13,15]. Figure 22.3 illustrates the proposed models of the fold surface in polymer lamellae [13,16].

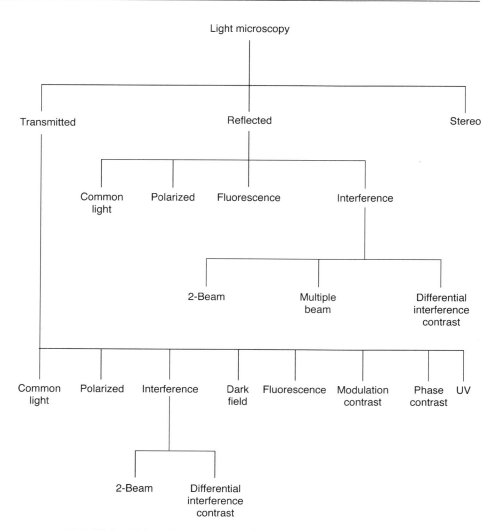

Fig. 22.1 Light microscopy techniques used to examine polymers.

The most prominent organization in polymers on a scale larger than lamellae is the spherulite, a spherical aggregate ranging from submicroscopic in size to millimeters in diameter [13]. Spherulites are recognized by their characteristic appearance in the polarizing microscope, where they are often seen as circular birefringent areas possessing a dark maltese cross pattern [13]. The birefringence effects are associated with molecular orientation resulting from the characteristic lamellar morphology. A typical polarizing optical micrograph of spherulites is shown in Fig. 22.4.

The structure [17–19] consists of radiating fibrils with amorphous material, additives, and impurities between fibrils and between individual spherulites. The crystalline part of the spherulite might be a lamellar or other crystalline structure.

Theory

In the OM, an image is produced by the interaction of light with the object of specimen [1]. The image can reveal fine detail in or on the specimen at a range of magnification from 2 to 2000X. Resolution on the order of 0.5 μm is possible, which is limited by the nature of the specimen, the objective lens, and the wavelength of light. Many textbooks deal with the subject at a variety of levels.

22. Optical Microscopy

Fig. 22.2 Lamellar structure of poly(ethylene).

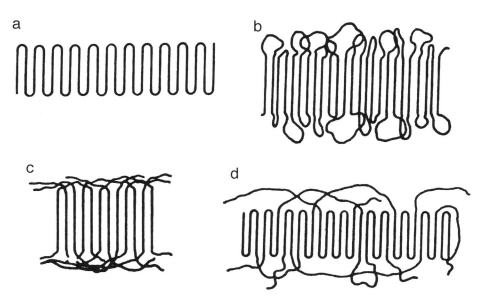

Fig. 22.3 Two-dimensional representations of models of the fold surface in polymer lamellae: (a) sharp folds, (b) loose loops, (c) "switchboard" model, and (d) a combination of all.

Fig. 22.4 Spherulitic structure of poly(amide) 12.

Spencer [20] deals with the fundamental science in a nonmathematical manner. Others are more comprehensive or have more practical information [21–27].

The information provided by the optical microscope concerns such things as the length, shape, relative arrangement, and orientation of visible features. Local measurement of optical constants such as refractive index and birefringence is also possible [1]. Many techniques are used to enhance contrast, which make more of the structure visible [1]. Optical microscopes have one or more light sources and a series of lenses that focus the light beam on the specimen and produce magnified images [1]. Images are typically recorded photographically or on videotape.

Imaging Modes

Bright Field

Bright field is the normal mode of operation of an optical microscope. The contrast in transmitted light is based on variations of optical density and color within the material. The absorption of light by pure polymer specimens is usually negligible at visible wavelengths. Carbon black agglomerates, pigment particles, and other fillers are clearly observed in polymers in bright field as the matrix polymers are typically transparent [1,2,28].

In reflected light, bright-field illumination is provided through the objective lens, perpendicular to the specimen. Resolution of polymer samples is often poor due to low surface reflectivity, scatter from within the specimen, and glare from other surfaces. A metal surface coating can be used to increase the surface reflectivity, which enhances brightness and contrast in reflected light.

Dark Field

A normal bright-field microscope can be modified in such a way as to exclude any undiffracted or unscattered light from the image-forming process [2]. The instrument is then operating in the "dark-field" mode [1,2,29]. Contrast in the image arises wherever there is a redistribution of the light passing through, or reflected from, the specimen. The background light intensity is theoretically zero and, as a result, a very high image contrast is available. Dark field in reflected light may be used to increase the contrast of surface roughness. If the specimen is not coated, dark field allows observation of subsurface features and details in reflected light.

Phase Contrast

Thin sections of polymer blends can give bright-field images with little or no contrast between the components. Transmitted light phase contrast converts the refractive index differences in such samples to light and dark image regions [1,2,30]. Small differences in thickness are also made visible [20–27]. Accessories for normal (Zernike) phase contrast include a special condenser with an opaque central plate and a matching phase ring in the back focal plane of the objective. Unscattered light passes through the phase ring and its phase is altered compared to the scattered light. Interference between scattered and direct rays causes changes in the image intensity. Light is scattered from the interfaces in a two-component transparent sample. The phase-contrast image typically has characteristic bright halos around fine structures due to some of the scattered light passing through the ring in the phase plate.

Interference Microscopy

In interference microscopy the illumination is split into two beams [1,2]. The beam splitter is a half-silvered mirror. In reflection, one beam is reflected from the sample, while the other is reflected from a flat reference mirror [20–22]. Transmission is more complex as the beam splitter may be a double refracting crystal, and the two beams can be displaced horizontally or vertically [20–22]. In all cases the two beams are recombined so that they interfere. The interference pattern can be used to measure the specimen thickness in transmission or the specimen roughness in reflection.

The plane of vibration of a linearly polarized beam emerging from a quarter-wave plate depends on the phase difference δ between the interfering ray and is measured by the angle α between the optical axes of the quarter-wave plate and the analyzer:

$$2\alpha = \delta (2\pi/\lambda)(n_2 - n_1)\ell,$$

where λ is the wavelength of the light, n_1 and n_2 are the refractive indexes of the sample and surrounding medium, and ℓ is the specimen thickness. If the refractive indexes are known, ℓ can be calculated.

Polarized Light

Polarized light microscopy is the study of the microstructures of objects using their interactions with polarized light [1,2,23–27,31–33]. The method is widely applicable to polymers [34] and to liquid crystals [34–37]. The polarizing microscope is basically a standard instrument fitted with a pair of polarizing filters. In the "crossed" position with their permitted vibration directions orthogonal, no light will pass through the microscope in the absence of a sample or if the sample is isotropic. Optically anisotropic, birefringent materials may appear bright between crossed polarizers [1,2]. When well-oriented specimens such as fibers are rotated on a rotatable stage, they go through four extinction positions of minimum intensity and four positions of maximum intensity [1,2]. In the extinction position the orientation direction is aligned parallel to one of the polarizer directions at 0° or 90°. Maximum intensity is at the 45° position. Circularly polarized light, obtained by addition of two crossed quarter-wave plates into the light path, is used to eliminate the extinction positions.

Birefringence

When a crystalline polymer is heated, the birefringence disappears gradually as the crystallites melt [12]. The point of disappearance of the last trace of birefringence, usually observed with the hot-stage microscope using crossed polarizers, is taken as the crystalline melting point. Any birefringence arising from strain or orientation in the amorphous regions disappears at the glass-transition temperature.

Birefringence is a useful property to monitor orientation [1,12,38]. Birefringence occurs as the molecular chains are aligned in the process of orientation; the magnitude and sign of the effect are determined by the chemical nature of the polymer.

Orientation

If orientation is assumed to occur only in one dimension (an oversimplification), birefringence and several related phenomena (infrared dichroism, etc.) measure the quantity $<\cos^2\theta>$, which is the average angle θ between the molecular chain direction and that of the orienting force, such as the fiber stretch axis. It is convenient to introduce an orientation function [12]

$$f = \tfrac{1}{2}(3<\cos^2\theta - 1>),$$

which has the value 0 for a completely unoriented or isotropic material for which $<\cos^2\theta> = \tfrac{1}{3}$, and 1 for a perfectly oriented fiber for which $<\cos^2\theta> = 1$. If all the molecular chains lie in the plane perpendicular to the fiber axis, $\theta = \pi/2$, then $<\cos^2\theta> = 0$ and $f = -\tfrac{1}{2}$. The orientation function is useful for relating the mechanical properties to the structure.

Small Angle Light Scattering

The polarizing microscope has been used to observe the small angle light scattering pattern (SALS) produced when polarized light passes through a specimen having a spherulite structure [40,41]. A typical crossed polar SALS pattern is shown in Fig. 22.5, which illustrates the formation of spherulites as a function of molecular weight and temperature. The separation of the diagonally opposite intensity maxima is inversely proportional to the spherulite size.

Measurement of Refractive Index

The refractive index, n, may be measured using an optical microscope [1,2,23,27,34]. Phase contrast increases the contrast due to differences in n and allows a more accurate determination. Interference contrast in transmission gives the optical path length and the average refractive index through the specimen thickness [1]. The Becke line method gives the surface refractive index [1].

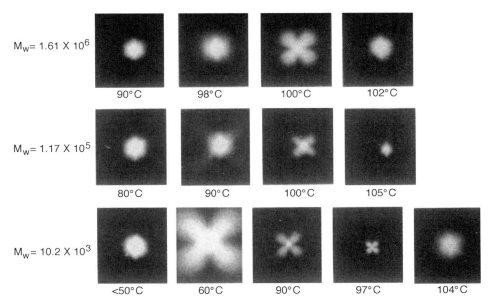

Fig. 22.5 A typical crossed polar SALS pattern.

Dynamic Microscopy

Most imaging of materials with microscopes is static, i.e., the sample is not intended to change during observation. Dynamic experiments can also be conducted in which the sample microstructure can be monitored as it changes as a function of temperature and thermal history [1,42,43]. Video cameras and recorders are the most widely used imaging devices for dynamic experiments [44].

There are two types of stages for dynamic microscopy: (a) hot and cold stages and (b) tensile stages [1]. Hot stages are most commonly used for the dynamic microscopy of polymers [1,43]. Thermal analysis in the OM is complementary to other thermal analysis methods, such as differential thermal analysis (DTA) [1]. Direct observation of the structural changes of a polymer as a function of temperature can determine the nature of phase changes and thermal decomposition [1]. It also measures the transformation temperatures.

Tensile stages are used to observe the deformation mechanisms of materials in tension [1,45]. A common use of tensile stages in the study of polymers is to study the tensile failure of fibers and yarns [45]. Other modes of deformation such as binding and shearing can be studied with suitably modified stages [1].

Sample Preparation

Certain types of samples require little preparation for light microscopy, whereas others require specific treatments and the use of special equipment. The methods of preparation employed depend not only on the physical form of the polymer, but also on its chemical and mechanical properties [1–4,27].

Small grains, powders, or large particle size latexes can be examined directly once dispersed on a microscopic slide and, should the optical system demand it, covered with a coverslip. Bulk samples can be prepared by a variety of techniques [2], as highlighted below.

Microtomy

The majority of synthetic polymers can be thin sectioned by microtomy for transmitted light purposes [2]. For optimum results, sectioning should be carried out at a temperature just below the glass/rubber transition temperature, T_g.

Embedding

Small samples, such as dry polymer powders, fibers, or thin films, need to be held for sectioning by embedding in a suitable carrier material. Waxes, epoxy, or acrylic resins are widely used for this purpose.

Staining Methods

In general, it is almost impossible to incorporate selective stains into polymers in a high enough concentration to produce contrast enhancement in thin sections [2]. An exception is the use of osmium tetroxide to stain unsaturated rubbers.

Melt Pressing

For transmitted light observation, thin layers of polymer can be obtained by pressing out small plates between the slide and the coverslip at elevated temperatures [2].

Solvent Casting

The "rule of thumb" for solvent casting is a 1% solution of polymer in a suitable solvent. One drop of the polymer solution is placed on a microscope slide, which is then heated on a hot plate to just above the boiling point of the solvent, which should produce a uniform film.

APPLICABILITY

The polarizing microscope can be used to observe morphological features of all crystalline polymers that can be prepared as thin films. Birefringence measurement is applicable to all polymers in which orientation or anisotropy can be induced.

ACCURACY AND PRECISION

Accuracy in determining the melting temperature, T_m, is difficult to specify as the melting point of any sample depends on its thermal history. The accuracy of the birefringence experiment in determining the sample thickness is probably no better than ±5% due to the limited control over the variables of the experiment.

The precision of the melting point determination using the hot stage depends on the heating rate. At 10°C/min, the T_m can be estimated to ±1°C. However, if the heating rate is 0.2°C/min, the precision in T_m is ±0.1°C.

SAFETY PRECAUTIONS

Safety glasses must be worn in the laboratory at all times. Appropriate gloves or other personal protective equipment may also need to be worn, depending on the polymer systems chosen for study.

Caution must be exercised around the illuminating source as it may be hot. Avoid looking directly at the illuminating source.

Observe all normal laboratory safety precautions during this experiment.

Material safety data sheets (MSDS) for all chemicals being used must be read prior to beginning the experiment. All chemicals should be considered hazardous from a standpoint of flammability and toxicity.

APPARATUS

1. Polarizing microscope equipped with cross-hair eyepiece, calibrating fine focus, and provision for insertion of compensater; a moderate magnification is desirable (e.g., 320× = 32× objective and 10× ocular)

22. Optical Microscopy

2. Compensator suitable for the microscope
3. Mechanical stage with vernier scales
4. Controlled temperature hot stage
5. Appropriate microscope illumination such as a mercury arc with monochromatizing filter for $\lambda = 546$ nm
6. Usual microscope accessories, including slides and cover glasses
7. Scissors or Exacto knife
8. Tweezers

REAGENTS AND MATERIALS

1. Linear poly(ethylene) (such as Marlex 50), poly(oxymethylene) (such as Delrin), or other highly crystalline polymer.
2. Poly(ethylene) wrapping, bag film, or other polymer that can be "cold drawn" without "necking down" in thin (0.05–0.2 nm) film form. Approximately 5 to 10 cm^2 is required.
3. Other semicrystalline polymers, such as nylon prepared in Experiment 9.

PREPARATION

Prepare the polymer samples for the cold drawing experiment by cutting six to eight specimens about 2 mm × 2 to 3 cm from the film described in the preceding section. Exact width is unimportant, but the samples must have a constant width and edges must be free from nicks or irregularities.

There are two simple ways to prepare samples: (a) with a very sharp razor blade or an Exacto knife, cut in one continuous motion and (b) using a sharp paper punch, cut two holes close together and cut two parallel lines to make "dumbbell-shaped" specimens.

PROCEDURE

Microscope Setup

Read the manufacturer's instructions prior to the start of the experiment.

1. Turn on the light source for the microscope approximately 10 min ahead of time to allow development of full intensity.
2. Insert the monochromatizing filter.
3. The sharpness and general character of the image are influenced by the illumination of the microscope. Depending on the age of the microscope used, this may involve removing the microscopic eyepiece, setting the polarizer and analyzer to 0°, and adjusting the position of the lamp and mirror while looking down the microscope tube to provide optimum, even illumination. Replace the ocular.
4. Make the necessary adjustments to obtain complete extinction, i.e., the darkest field possible.
5. If a recorder is attached to the microscope, the light intensity must be within the recorder sensitivity, both with and without the sample.
6. Place the hot stage unit onto the microscope.

Crystallization and Melting of Linear Poly(ethylene) [46]

1. Cut a small piece of poly(ethylene), no more than 0.5 nm on a side.
2. Place the cut sample on a clean, dust-free microscope slide in such a way as to ensure that after the slide is inserted into the hot stage, the sample comes into the field of view.

3. Put a clean, dust-free cover glass over the sample.
4. Place the slide onto the hot stage unit.
5. Clamp the slide in place.
6. Carefully lower the microscopic objective until it almost touches the slide on the hot stage.
7. Focus on the sample by raising the objective. *Warning:* Do not lower the objective as it will crack the sample slide and possibly damage the objective.
8. Adjust the condenser focus and diaphragm for even illumination at a convenient level.
9. Turn on the hot stage unit.
10. Set the temperature of the hot stage to 30°C above the melting point, T_m, of the polymer. [For linear poly(ethylene), the hot stage temperature should be set to 65°C] (see Note 1).
11. Allow the sample to equilibrate for 3 min (see Note 2).
12. Cool the sample rapidly to about 40°C below T_m. [For linear poly(ethylene), this temperature would be 95°C.]
13. Watch for the onset of crystallization, which may occur even before the temperature equilibration is achieved.
14. When the crystallization appears to be complete, heat the sample at an appropriate rate (e.g., 10°C/min).
15. Determine the melting point of the polymer as follows:
    a. When the sample starts to melt, record the temperature as T_A.
    b. When the light intensity is reduced to one-half, record this temperature as T_B.
    c. When the field becomes completely dark, record this temperature as T_C.
    d. Record these temperatures (see Notes 2 and 3).
16. Repeat steps 10 to 15, but on step 12 in successive experiments, vary the temperature in the following way: 115°, 120°, 122°, and 124°C (see Note 4).
17. If time permits, determine T_m for other crystalline polymers by repeating steps 10 to 16.
18. Finally, turn the hot plate control off and remove it from the microscope.

Birefringence

1. Insert the compensator into the slot on the barrel of the microscope.
2. Mount the mechanical stage on the microscope.
3. Using a fine tip pen or marker, mark the polymer strips (see Preparation) with two lines approximately 1 mm apart in order to determine the elongation.
4. Place the sample on a clean, dust-free microscope slide mounted on the mechanical stage of the microscope, aligned so that the ink marks are parallel to one axis of the stage.
5. Focus on the center of the sample (between the ink marks) so that both the sample and the surrounding field can be seen.
6. With the compensator set for no compensation and (depending on the age of the microscope) and the polarizer and analyzer set for extinction, rotate the microscope stage, observing the intensity of light transmitted through the sample. A point of minimum intensity should easily be found (see Note 5).

7. Set the stage to the position of the minimum.
8. Read the position using the scale and vernier on its circumference.
9. Record the value.
10. Set the stage exactly 45° from the position found in step 6.
11. Turn the compensator drum until the sample and field show equal brightness.
12. Using the vernier, record the compensator drum reading to the nearest 0.1°.
13. Return the compensator drum to its original position and rotate it in the opposite direction until a second position of equal brightness is found.
14. Record this angle as well as half the difference between the two readings as the angle θ.
15. Shift the mechanical stage until each of the ink marks coincides with the ocular cross hair.
16. Record both positions by reading the mechanical stage scale and vernier.
17. Subtract the values of the two positions to find the length, L, of the sample. Record the value.
18. Measure the sample thickness by focusing carefully on its upper, then lower, surface.
19. At each point, record the setting of the fine-focus knob.
20. The difference in the readings is the sample thickness, ℓ. Record this size.
21. Remove the sample from the microscope slide.
22. Using a pair of tweezers, stretch the polymer slowly and evenly (see Note 6). Do not exceed a 20% elongation.
23. Repeat steps 4 to 22 for several stages of elongation of the sample. Keep the elongations small to avoid exceeding the range of the compensator.
24. Remove the sample and the mechanical stage.
25. Remove the compensator from the microscope.
26. Turn off the microscope lamp.

FUNDAMENTAL EQUATIONS

Percent Elongation

$$E = 100\ (L-L_0)/L_0,$$

where L_0 and L are the initial and final lengths of the sample at each stage of elongation.

Retardation

$$\Gamma = C\ f(\theta)$$

$$\delta = \Gamma\ (2\pi/\lambda),$$

where C for $\lambda = 546$ nm and tables of $f(\theta)$ are supplied with the compensator.

Birefringence

$$\Delta n = \lambda\delta/2\pi\ell = \Gamma/\ell$$

CALCULATIONS

Using the equations just given, calculate E, Γ, δ, and Δn for each elongation of the polymer sample.

REPORT

1. Describe the equipment and objective of the experiment.
2. Describe the morphology of the polymer sample as observed at each crystallization temperature. Comment on any difference.
3. Tabulate crystallization temperatures and melting temperature range, T_A, T_B, and T_C, for the polymer sample. Plot T_A, T_B and T_C versus crystallization temperature and include the graph.
4. Tablulate L or (L_0), E, ℓ, θ, Γ, δ, and Δn.
5. Plot Δn versus E and include the graph in the report.

NOTES

1. At the most rapid heating or cooling rate, the sample temperature lags behind the indicated temperature.
2. If the system is automated, temperature profiles/jumps as well as equilibration times can be programmed.
3. If the system has a photomonitor or equivalent, the melting process can be recorded in photographs.
4. At higher crystallization temperatures, wait at least 15 min for the completion of crystallization.
5. If a point of minimum intensity cannot by found, the sample must be annealed using the hot stage to reduce its anisotropy.
6. Care and some practice are required to avoid breaking the sample.

ACKNOWLEDGMENTS

The authors thank Drs. Marina Despotopoulou and Kent Zhang, Elf Atochem North America, for proofreading and making corrections. The authors also thank Elf Atochem SA for some of the figures.

REFERENCES

1. L. C. Sawyer and D. T. Grubb, "Polymer Microscopy," Chapman and Hall, New York, 1987.
2. D. Hemsley, "Comprehensive Polymer Science: The Synthesis, Characterization, Reactions and Applications of Polymers" (G. Allen and J. C. Bevington, eds.), Vol. 1, Chap. 33, p. 765, Pergamon Press, New York, 1989.
3. S. Y. Hobbs, *J. Macromol. Sci. Rev. Macromol. Chem.* **C19**(2), 221 (1980).
4. S. Y. Hobbs, in "Plastics Polymer Science and Technology" (M. D. Bayal, ed.), p. 239, Wiley-Interscience, New York, 1982.
5. P. H. Geil, "Polymer Single Crystals," Interscience, New York, 1963.
6. A. Keller, *Rep. Prog. Phys.* **31**, 623 (1968).
7. B. Wunderlich, "Macromolecular Physics," Vols. 1 and 2, Academic Press, New York, 1973.
8. D. T. Grubb, "Developments in Crystalline Polymers," Vol. 1, Applied Science, London, 1982.
9. D. C. Bassett, "Principles of Polymer Morphology," Cambridge University Press, Cambridge, 1981.
10. R. B. Seymour, ed., "The History of Polymer Science and Technology," Dekker, New York, 1982.
11. H. D. Keith, *Kolloid Z. Z. Polym.* **231**, 421 (1969).

12. E. A. Collins, J. Bareŝ, and F. W. Billmeyer, Jr., "Experiments in Polymer Science," Wiley-Interscience, New York, 1973.
13. F. W. Billmeyer, Jr., "Textbook of Polymer Science," 3rd Ed., Wiley-Interscience, New York, 1984.
14. P. J. Hendra, in "Structural Studies of Macromolecules by Spectroscopic Methods" (K. J. Ivin, ed.), Chap. 6, Wiley, New York, 1976.
15. A. Keller, *Philos Mag.* **2**(8), 1171 (1957).
16. P. Ingram and A. Peterlin, in "Encyclopedia of Polymer Science and Technology" (H. F. Mark, N. G. Gaylord, and N. M. Bikalis, eds.), Vol. 9, p. 204, Wiley-Interscience, New York, 1968.
17. H. D. Keith and F. J. Padden, *J. Appl. Phys.* **34,** 2409 (1963).
18. I. M. Ward, ed., "Structure and Properties of Oriented Polymers," Applied Science, London, 1975.
19. A. Peterlin, in "Structure and Properties of Oriented Polymers" (I. M. Ward, ed.), p 36, Wiley, New York, 1975.
20. M. Spencer, "Fundamentals in Light Microscopy," Cambridge University Press, Cambridge, 1982.
21. W. G. Hartley, "Hartley Microscopy," Senecio, Charlbury, 1981.
22. S. Bradbury, "An Introduction to the Optical Microscope," Oxford University Press, Oxford, 1984.
23. F. A. Jenkins and H. E. White, "Fundamentals of Optics," 4th Ed., McGraw Hill, New York, 1976.
24. R. B. McLaughlin, "Special Methods in Light Microscopy," in The Microscopic Series, Vol. 17, Microscopic Publications, London, 1977.
25. J. H. Richardson, "Optical Microscopy for Materials Sciences," Dekker, New York, 1971.
26. G. L. Clark, ed., "The Encyclopedia of Microscopy," Reinhold, New York, 1961.
27. C. W. Mason, "Handbook of Chemical Microscopy," 4th Ed., Wiley, New York, 1983.
28. J. W. Ess, P. R. Hornsby, S. Y. Lin, and M. J. Brevis, *Plast. Rubber Proc. Appl.* **4,** 17 (1984).
29. B. Mutagahyiva and D. A. Hemsley, *Plast. Rubber Proc. Appl.* **5,** 3 (1985).
30. D. J. Goldstein, *J. Microsc. (Oxford)* **128,** 33 (1982).
31. E. A. Wood, "Crystals and Light," 2nd Ed., Dover, New York, 1977.
32. P. F. Kerr, "Optical Microscopy," McGraw Hill, New York, 1959.
33. P. Gay, "An Introduction to Crystal Optics," Longmans, London, 1967.
34. N. H. Hartshorne and A. Stuart, "Crystals and the Polarizing Microscope," Arnold, London, 1970.
35. G. Friedel, *Ann. Phys.* **18,** 273 (1922).
36. N. H. Hartshorne, "The Microscopy of Liquid Crystals," Microscope Publications, Chicago, 1974.
37. A. F. Hallimond, "The Polarizing Microscope," 3rd Ed., Vickers, York, 1970.
38. G. I. Wilkes, *Adv. Polym. Sci.* **8,** 91 (1971).
39. H. Kawai and S. Nomura, in "Developments in Polymer Characterization" (J. V. Dawkins, ed.), Applied Science, London, 1983.
40. D. A. Hemsley, "The Light Microscopy of Synthetic Polymers," Oxford University Press, Oxford, 1984.
41. R. S. Stein and M. B. Rhodes, *J. Appl. Phys.* **31,** 1873 (1960).
42. E. P. Butler and K. F. Hale, in "Practical Methods in Electron Microscopy," (A. M. Glauert, ed.), North-Holland, Amsterdam, 1981.
43. J. A. Reffner, *Am. Lab.,* April 1984, p. 29.
44. K. F. Hale and M. H. Brown, *Micron.* **4,** 434 (1973).
45. J. W. S. Hearle, J. T. Sparrow, and P. M. Cross, "The Use of the Scanning Electron Microscope," Pergamon Press, Oxford, 1972.
46. ASTM F 766-82.

EXPERIMENT 23

Dynamic Mechanical Analysis

INTRODUCTION

Polymers vary from liquids and soft rubbers to very hard and rigid solids. Many structural factors determine the nature of the mechanical behavior of such materials. In considering structure–property relationships, polymers may be classified into one of several regimes, shown in the volume–temperature plot (Fig. 23.1).

Dynamic mechanical analysis (DMA) or dynamic mechanical thermal analysis (DMTA) provides a method for determining elastic and loss moduli of polymers as a function of temperature, frequency or time, or both [1–13]. Viscoelasticity describes the time-dependent mechanical properties of polymers, which in limiting cases can behave as either elastic solids or viscous liquids (Fig. 23.2). Knowledge of the viscoelastic behavior of polymers and its relation to molecular structure is essential in the understanding of both processing and end-use properties.

DMA can be applied to a wide range of materials using the different sample fixture configurations and deformation modes (Table 23.1) [10,11]. This procedure can be used to evaluate by comparison to known materials: (a) degree of phase separation in multicomponent systems; (b) amount type, and dispersion of filler; (c) degree of polymer crystallinity, (d) effects of certain pretreatment; and (e) stiffness of polymer composites [8,11].

PRINCIPLE

The Modulus Curve

Dynamic mechanical experiments yield both the elastic modulus of the material and its mechanical damping, or energy dissipation, characteristics. These properties can be determined as a function of frequency (time) and temperature. Application of the time–temperature equivalence principle [1–3] yields master curves like those in Fig. 23.2. The five regions described in the curve are typical of polymer viscoelastic behavior.

In the glassy region, the polymer is below its glass transition temperature, T_g, and typically has a modulus of 10^{10} dynes/cm^2. The transition region includes the T_g, which is taken as the point of inflection of the modulus or the maximum in the damping curve. The modulus drops by a factor of 1000 in this region. The

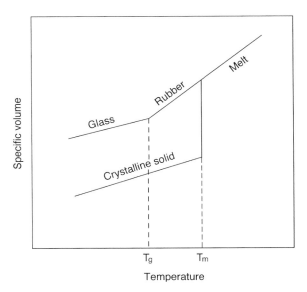

Fig. 23.1 Regimes of bulk polymers in terms of volume and temperature.

third region is termed the rubbery plateau. The modulus varies little in this region. The fourth region is that of elastic or rubbery flow, following liquid flow in which there is very little elastic recovery. In this region the modulus may fall below 10^5 dynes/cm^2.

Theory

The sample is clamped into a frame in the DMA experiment. An applied sinusoidally varying stress is applied to a polymer sample at a frequency ω with a small amplitude, which can be represented as [1,2,8,10–13]

$$\sigma(t) = \sigma_0(t) + (\omega t + \delta), \qquad (1)$$

where σ is the maximum stress amplitude, t is time, ω is angular frequency, and the stress precedes the strain by a phase angle δ. The strain is given by

$$\gamma(t) = \gamma_0 \sin(\omega t), \qquad (2)$$

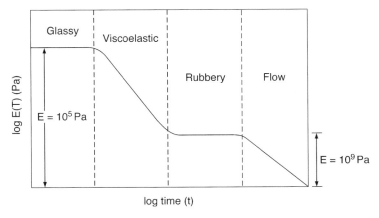

Fig. 23.2 A plot of the stress relaxation modulus as a function of time.

TABLE 23.1
DMA Deformation Modes for Specific Applications

Sample	Parameter	Clamp/deformation mode
Solid polymer	Dynamic modulus	Flexure
	Glass transition temperature	Tension
	Melting temperature	Torsion
	Cross-link density	Compression
	Relaxation behavior	Shear
	Crystallinity, cure	
Film, fiber, coatings	Dynamic modulus	Flexure
	Glass transition temperature	Tension
	Creep, cure, compliance	Shear
	Relaxation behavior	
Viscous fluids, gels	Viscosity	Shear
	Gelation	
	Gel–Sol transition	
	Cure, dynamic modulus	

where γ_0 is the maximum strain amplitude (Fig. 23.3). Expanding Eq. (1) gives

$$\sigma(t) = \sigma \sin \omega \cos \delta + \sigma_0 \sin \omega\, t \sin \delta. \qquad (3)$$

The peak stress can be resolved into a component $\sigma_0 \cos \delta$ that is in place with the strain, related to the stored elastic energy and a component $\sigma_0 \sin \delta$, which is 90° out of phase with the strain, related to the viscous loss of energy [1,3,9,10–13].

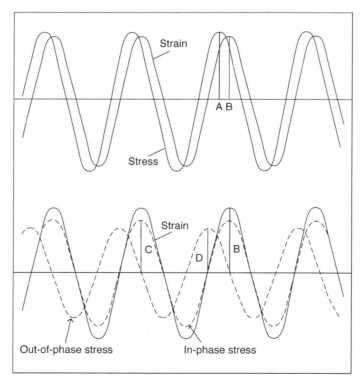

Fig. 23.3 Sinusoidally varying stress and strain in a dynamic mechanical experiment.

The stress and strain are related by [1,2,8,10–13]

$$\sigma(t) = E_0^*(\omega)\delta(t), \tag{4}$$

where $E^*(\omega)$ is the dynamic modulus, and

$$E^*(\omega) = E'(\omega) + iE''(\omega), \tag{5}$$

where $E'(\omega)$ and $E''(\omega)$ are the dynamic storage modulus and the dynamic loss modulus, respectively (Fig. 23.3). For a viscoelastic polymer, E' characterizes the ability of the polymer to store energy (elastic behavior), whereas E'' reveals the tendency of the material to dissipate energy (viscous behavior).

The modulus can be defined [1,2,5,8,10–13] as

$$E' = (\sigma_0/\gamma_0)\cos\delta = \frac{C}{B} \tag{6}$$

and

$$E'' = (\sigma_0/\gamma_0)\sin\delta = \frac{D}{B}. \tag{7}$$

The ratio of the two is the dissipation factor, tan δ, which is an indicator of the relative importance of the viscous as compared to the elastic aspects of the polymer's behavior [1,2,9–13]:

$$\tan\delta = E''/E' = A/B \tag{8}$$

Normally, E', E'', and tan δ are plotted against temperature or time (Fig. 23.3).

Amorphous Polymers

In the glassy state, the amorphous chains are frozen into a rigid disordered structure, giving a high modulus and a low loss factor [1]. Some limited mobility is possible, which gives rise to one or more transitions of low magnitude. The relaxation transitions are usually labeled in alphabetical order, α, β, γ, and δ, with decreasing temperature, with the highest transition, the α relaxation, being the glass-to-rubber transition (Fig. 23.4) [18]. The glass-to-rubber transition

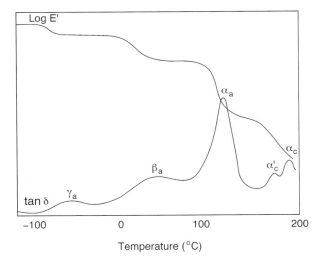

Fig. 23.4 Typical transition behavior in mechanical storage modulus for a semicrystalline polymer. The subscripts a and c refer to the amorphous and crystalline phases, of the polymer, with α_a as the main T_g process.

changes the properties of amorphous polymers from those of a rigid, essentially elastic glass to those of a flexible, high loss rubber. The temperature at which this transition occurs will depend on a number of factors, such as molecular weight, chain flexibility, and plasticizers. The β transition arises from the motion of side groups on the polymer backbone and often indicates a certain ability to absorb impact in the glassy region [14].

Chain Flexibility

Chains containing mainly methylene sequences or methylene sequences plus linkages have a higher chain flexibility and a low glass transition temperature [1,8,15]. Polymer chains containing rigid phenylene groups, such as poly(phenylene oxide), have a low chain flexibility and a high glass transition temperature [1].

Side groups also influence chain flexibility. As the size and the rigidity of the side groups increase, the chain flexibility is lowered and the glass transition temperature is increased [16]. In contrast, introduction of lengthy flexible nonpolar side groups decreases T_g [15,17].

Cross-linking

Cross-linking has a dramatic effect on the dynamic mechanical properties above the non-cross-linked polymer T_g. At extremely high cross-linking, the glass transition either disappears or is shifted to such a high temperature that the polymer decomposes before reaching T_g [1,8,15]. It is generally assumed that the broadening of the transition is due to a distribution of the molecular weight between the cross-links or some other kind of heterogeneity in the network structure [18]. In contrast, Mason [19] believes that the broadening is due to a broadening in the high distribution of free volume between monomeric units.

Plasticizers

Plasticizers are liquids that are added to plastics to soften them [1,8,15]. The softening is brought about by the plasticizer dissolving in the polymer and lowering its glass transition temperature [15].

Fillers

Fillers are frequently added to polymers to lower their cost [1]. As a result of the presence of filler, the mechanical properties of the material are changed [1]. Impact and elongation are usually decreased whereas tensile strength, hardness and stiffness are increased [1].

If polymer molecules are adsorbed on the filler surface their mobility is restricted [1]. As a result, the glass transition of the adsorbed polymer will increase [1]. Often two transitions can be seen, corresponding to the glass transition of the unadsorbed and adsorbed polymer [20].

Crystalline Polymers

The assignment of relaxation transitions to molecular or structural processes in crystalline polymer is not as well established as for amorphous plastics [1]. The following types of relaxations can be identified: (a) in the amorphous phase, (b) in the crystalline phase, (c) in both the crystalline and the amorphous phases, and (d) those involving large-scale features of the crystalline morphology [1]. Relaxations in the amorphous phase are reflected in the glass transition, T_g. Relaxations within the crystalline phase could be the movements of defects or cooperative movement of chains. Relaxations involving large-scale features of the crystalline morphology might involve shear deformations between lamellae or fibrils [1,21].

Elastomers

The dynamic mechanical properties of elastomers have been extensively studied since the mid-1940s by rubber physicists [1]. Elastomers appear to exhibit extremely complex behavior, having time–temperature- and strain–history-dependent hyperelastic properties [1]. As in polymer cures, DMA can estimate the point of critical entanglement or the gel point.

APPLICABILITY

The dynamic mechanical behavior of most homogeneous and heterogeneous solid and molten polymeric systems or composite formulations can be determined by DMA. These polymeric systems may contain chemical additives, including fillers, reinforcements, stabilizers, plasticizers, flame retardants, impact modifiers, processing aids, and other chemical additives, which are added to the polymeric system to impart specific functional properties and which could affect the processability and performance.

ACCURACY AND PRECISION

It is difficult to specify accuracy in this experiment. One reason is that there may be sampling effects, i.e., wide variability in the samples used. Consequently, the sample should be homogeneous and representative. There is a strong dependence of the modulus and damping behavior on molecular and structural parameters. Entrapped air/gas may affect the results obtained using powder or pellet samples.

SAFETY PRECAUTIONS

Safety glasses must be worn at all times in the laboratory. Other personal protection equipment, such as appropriate gloves and/or a lab coat, should also be worn.

The instrument should be in a well-ventilated area, especially at experiments done above room temperature as the polymer or the additives may be subject to decomposition/degradation phenomena.

Care should be taken during variable temperature experiments. When using liquid nitrogen for variable temperature work, skin contact should be avoided as severe frostbite or cryogenic burns can result. Also, certain parts of the equipment may become hot during the experiment.

All normal laboratory safety precautions should be observed. Material safety data sheets (MSDS) for all chemicals being used must be read prior to beginning the experiment. All chemicals should be considered hazardous from a standpoint of flammability and toxicity.

APPARATUS

1. DMA Equipment
   a. Appropriate test fixtures
   b. Temperature programmer/controller
   c. Sample holder and cooling attachments if needed for subambient testing

d. Data collection and output devices such as X-Y recorder or chart recorder or computer analysis system with plotter
2. Sample preparation tools to size sampler for sample holder
    a. Heatable platen press
    b. Cutting tools as needed

REAGENTS AND MATERIALS

1. Polymers
    a. Synthesized polymers from earlier experiments
    b. Commercial polymers: semicrystalline polymers such as poly(ethylene) or poly(ethylene terephthalate); amorphous polymers such as plasticized poly(vinyl chloride) (flexible at ambient temperature).
2. Nitrogen or other gas supply for purging purposes

PREPARATION

1. Prepare the sample, if not readily available, in the form of a film approximately 1 mm thick or less for rigid polymers or 4 mm thick for flexible polymers. Film preparation methods include casting from solution, milling, or compression molding (Note 1).
2. Cut a specimen 3–5 mm wide and about 15 mm long (for elevated temperature experiments) or 25 mm long (for subambient temperature measurements). Use a single stroke with a sharp razor blade, avoiding nicks or rough edges on the sides of the sample.

PROCEDURE

The manufacturer's information on the instrument should be read prior to the start of the experiment. It is also recommended that References 2,11,12, and 21–24 be read.

Calibration

a. Calibrate the instrument using the procedure recommended by the manufacturer.

Measurement

Use an untested specimen for each measurement.

a. Measure the diameter and height (thickness) of the specimen to the nearest 0.03 mm (0.001 in.) at the center of the specimen. Record the measurement.
b. Compress the test specimen between movable and stationary members.
c. Preload the sample so that there is a positive force. Monitor the normal force to ensure adequate preloading.
d. Measure the jaw separation between movable and stationary flat plates to the nearest 0.03 mm (0.001 in.). Record the measurement.

e. If variable temperature is used, then this measurement must be corrected for the thermal expansion of the fixtures during testing.
f. Select the desired frequency (or frequencies) for dynamic linear displacement.
g. Select the linear displacement amplitude.
h. Temperature increases should be controlled to 1 to 2°C/min for linear increases and 2 to 5°C/min, with a minimum of 3-min thermal soak time, for step increases. This will allow characterizing of the modulus.

FUNDAMENTAL EQUATIONS

$$E' = E^* \cos \delta$$
$$E'' = E^* \sin \delta = E' \tan \delta$$

CALCULATIONS

Process collected data using the instrument data analysis system and plot the calculated values of storage (elastic) modulus (E'), loss (viscous) modulus (E''), complex modulus (E^*), and tanδ versus temperature.

REPORT

a. Describe the instrument used for the test.
b. Describe the instrument calibration procedure.
c. Describe the experiment performed. Include a description of (1) the fixtures used, (2) the sample and its preparation prior to test, (3) the gaseous atmosphere used during the test, and (4) the frequency or frequency range used.
d. Tabulate data and results, including the moduli complex viscosity, and tan δ as a function of the dynamic oscillation (frequency), percent strain, temperature, or time.
e. Report the value of the glass transition temperature if this region was investigated.
f. Comment on the significance of tan δ.
g. Comment on the elastic and viscous nature of the material at different temperatures.

NOTES

1. Refer to Ref. [25] for guidelines to prepare compression molded plaques, sheet or test specimens.
2. A residual solvent from casting may affect the results and milling requires a large sample.

ACKNOWLEDGMENTS

The authors thank Dr. Larry Judovits, Elf Atochem North America, Kathy Lynn Lavanga, Rheometric Scientific, and Dr. James S. Holton for proofreading and providing comments and corrections.

REFERENCES

1. P. Gradin, P. G. Howgate, R. Seldén, and R. A. Brown, "Comprehensive Polymer Science: The Synthesis, Characterization, Reactions and Application of Polymers" (G. Allen and J. C. Bevington, eds.), Vol. 1, p. 533, Pergamon Press, New York, 1989.
2. E. A. Collins, J. Bareŝ, and F. W. Billmeyer, Jr., "Experiments in Polymer Science," Wiley, New York, 1973.
3. F. W. Billmeyer, Jr., "Textbook of Polymer Science," 3rd Ed., Wiley, New York, 1984.
4. T. Alfrey, Jr., "Mechanical Behavior of High Polymers," Interscience, New York, 1948.
5. J. D. Ferry, "Viscoelastic Properties of Polymers," 2nd Ed., Wiley, New York, 1970.
6. A. V. Tabolsky, "Properties and Structure of Polymers," Wiley, New York, 1960.
7. J. V. Schmitz, ed., "Testing of Polymers," 4 Volumes, Interscience, New York, 1965, and later years.
8. L. E. Nielsen, "Mechanical Properties of Polymers and Composites," Vol. 1, Dekker, New York, 1974.
9. R. E. Wetton, in "Polymer Characterization" (B. J. Hart and M. I. James, eds.), Blackie Academic and Professional, New York, 1993.
10. T. Hatakeyama and F. X. Quinn, "Thermal Analysis: Fundamentals and Application to Polymer Science," Wiley, New York, 1994.
11. ASTM D 4065-95, "Standard Practice for Determining and Reporting Dynamic Mechanical Properties of Plastics," ASTM, West Conshohocken, PA, 1997.
12. ASTM D 4440-95a, "Standard Practice for Rheological Measurement of Polymer Melts Using Dynamic Mechanical Procedures," ASTM, West Conshohocken, PA, 1997.
13. M. E. Brown, "Introduction to Thermal Analysis: Techniques and Applications," Chapman and Hall, New York, 1988.
14. T. G. Fox and P. J. Flory, *J. Polym. Sci.* **14,** 315 (1954).
15. D. J. Williams, "Polymer Science and Engineering," Chap. 7, p. 209, Prentice-Hall, New Jersey, 1971.
16. T. G. Fox and P. J. Flory, *J. Appl. Phys.* **21,** 581, (1950).
17. P. I. Vincent, in "Physics of Plastics" (P. D. Ritchie, ed.), Iliffe Books, London, 1965.
18. K. Ueberreiter and G. Kanig, *J. Chem. Phys.* **18,** 399 (1950).
19. P. Mason, *Polymer* **5,** 625 (1964).
20. A. Yim, R. S. Chakal, and L. E. St. Pierre, *J. Colloid. Interface Sci.* **43,** 583 (1973).
21. ASTM E 1640-94, "Standard Test Method for Assignment of the Glass Transition Temperature by Dynamic Mechanical Analysis," ASTM, West Conshohocken, PA, 1994.
22. ASTM D 5026-95a, "Standard Test Method for Measuring the Dynamic Mechanical Properties of Plastics in Tension," ASTM, West Conshohocken, PA, 1997.
23. ASTM D 2023-95a, "Standard Test Method for Measuring the Dynamic Mechanical Properties of Plastics Using Three Point Bending," ASTM, West Conshohocken, PA, 1997.
24. ASTM D 5024-95a, "Standard Test Method for Measuring the Dynamic Mechanical Properties of Plastics in Compression," ASTM, West Conshohocken, PA, 1997.
25. ASTM D 4703-93, "Standard Practice for Compression Molding Thermoplastic Materials into Test Specimens, Plaques, or Sheets," ASTM, West Conshohocken, PA, 1997.

Index

Acid catalysts,
   for cationic polymerization, 22
Acrylic esters,
   polymerization, 27ff
Alcoholysis,
   of poly(vinyl acetate), 77–80
Aluminum Alkyls, 23
Anionic polymerization,
   of alkylisocyanates, 46
   of styrene, 17–21
Aramide, 47

Beer's Law, 102, 105
Benzoin, 35–36
Birefringence, 123, 186, 190, 192, 195
   orientation, 190
Bisphenol A Diglycidyl Ether,
   curing of, 69–70
Bisphenols, 61
Bragg's law, 175, 176, 181, 183
Brownian motion, 152
Bulk polymerization,
   of acrylic esters, 29–30, 35
   of styrene, 10–12
Butyl acrylate,
   in emulsion terpolymerization, 73–76
N-Butyllithium, 20

Carbonium ions, 22
Cationic Polymerization,
   of styrene, 22ff
Cellulose, 131, 133
Chain flexibility, 202
Chain Transfer Agents, 28, 31
Copolymerization,
   of predetermined composition, 74

"Core-shell" particles, 74
Crosslinking, 202
   poly(vinyl acetate), 77–78
Crystals, single, 185
   lamellae, 185
   spherulites, 185, 186
Crystallinity,
   by nuclear magnetic resonance,
     (NMR), 86, 90
   by differential scanning calorimetry,
     (DSC), 120
   by dynamic mechanical analysis,
     (DMA), 198
   by infrared spectroscopy, (IR), 101
   by X-ray diffraction, (XRD), 173
Curing, 103
   dynamic mechanical analysis, (DMA),
     203
   infrared spectroscopy, (IR), 103

Degradation, polymer,
   random chain scission, 110
   systematic chain scission, 110
Differential Scanning Calorimetry,
   (DSC), 120–130
   crystallization temperature, T_c, 120
   decomposition temperature, 120, 124
   enthalpy, 120, 122–124
   entropy of fusion, ΔS, 123
   glass transition temperature, T_g, 120
   heat capacity, C_p, 122, 123
   heat of fusion, ΔH, 122
   melt temperature, T_m, 120, 123, 124
   specific heat, c_p, 122, 123
   thermal conductivity, 121, 122
Differential thermal analysis, (DTA), 120

Diffraction measurements, 185
   electron, 185
   2,4-dinitrobenzenesulfonyl chloride, 165
Dispersion polymerization, 30
DNA, 133
Dynamic Mechanical Analysis, (DMA), 198–206
   chain flexibility, 202
   composites, 198
   crosslinking, 202
   crystallinity, 198
   dissipation factor, tan δ, 201
   elastic modulus, 198
   energy dissipation, 198
   fillers, 202, 203
   glass-to-rubber transition, 201
   glassy region, 198
   glass transition temperature, T_g, 198
   loss modulus, 198
   mechanical damping, 198
   plasticizer, 202
   rubber, 199
   strain, 200
   stress, 200
   tensile strength, 202
   viscoelasticity, 198
Dynamic Mechanical Thermal Analysis, (DMTA), 198–206

Elastomers, 203
End group analysis, 165
Emulsion Polymerization,
   of acrylic esters, 27, 30, 31–32
   of ethyl acrylate, 41ff
   redox, 41
   seeded terpolymerization, 73–76
   of styrene, 14
End Group Analysis, 163–172
   acetylation, 165
   branching, extent of, 164
   bromination, 165
   carbonyl groups, 165
   epoxy groups, 165
   equivalent weight, 166
   kinetics, 163
   mercapto groups, 165
   molecular weight, number average, M_n, 163, 166
   olefinic bouble bonds, 165
   phenolic hydroxyl groups, 165
   phthalation, 165
   repeat units, 166
Epichlorohydrin, 61, 63

Epoxy compounds,
   curing of, 64–67, 69
   preparation of, 63–64
Epoxy resins, 61
Epoxy value determination, 63
Ethyl Acrylate,
   emulsion polymerization, 41ff

Fibers, 176, 177
Fibrils, 203
Flame retardants, 203
Flexible chains, 133
Free Radical Initiators, 86
Free Radical Polymerization,
   of acrylic esters, 27ff
   monomer properties on procedures, effect of, 71–72
   of styrene, 9–16
   of vinyl acetate, 71
   of vinyl monomers, 27

Gel Permeation Chromatography, (GPC), 140–151
   detectors, 144–145
   eluents, 144
   hydrodynamic volume, 141–143, 145
   plate height, 143
   plate number, 143
Glass transition temperature, T_g, 120.
Glycolide, 56
GMP, *see* Good Manufacturing Practices
Good Manufacturing Practices, 2

Heat of polymerization,
   of acrylic esters, 29
Heat capacity, C_p, 122, 123
Heat of fusion, 122, 123
Hydrodynamic volume, 141, 145

Impact modifiers, 203
Infrared Spectroscopy, 98–107
   absorbance, A, 102
   absorptivity, molar, ε, 102, 103
   attenuated total reflectance, (ATR), 104
   Beer's Law, 102, 105
   crystallinity, 101, 103
   curing, 103
   dipole moment, 100
   far-infrared, 98, 99

fingerprint region, 98
group frequency region, 98
internal reflectance, 104
kinetics, polymerization, 103
microscopy, (IR), 104
microstructure, 101
mid-infrared, 98, 99
molecular orientation, 103
molecular weight average, 103
near-infrared, (NIR), 98, 99
orientation, 101
rotational energy, 100
thermal degradation, 103
transmittance, T, 102
vibrational energy, 100
Inhibitors, 27–28, 36
Initiators,
 acrylic ester polymerization, 28–29, 31, 32, 35, 42
 for epoxy resin curing, 65
 styrene polymerization, 9, 15, 17–19, 22, 24
 vinyl acetate polymerzation, 75
Interfacial polycondensation, 45–46, 49–51
Isoprene rubber, 175

Kevlar, 47
Kinetics
 by end group analysis, 163
 by infrared analysis, 103
 by nuclear magnetic resonance, 92–94

Lamellae, 185, 203
 by optical microscopy, 185
 by X-ray diffraction, 173
Light Scattering, 144–146, 152–162, 190
 Brillouin scattering, 152
 Brownian movement, 152
 elastic scattering, 152
 low angle, laser light scattering, (LALLS), 145
 nonelastic scattering, 152
 quasielastic scattering, 152
 Raman scattering, 152
 Rayleigh factor, R_θ, 145
 Rayleigh line broadening, 152
 small angle (SALS), 190
 thermodynamic quantities, 152
Liquid crystals, 54, 55, 189

"Living" polymers, 18
Log of Operation, 3, 4

Mark–Houwink Equation, 133, 134
Mark–Houwink–Sakurada constant, 142
Marlex, 179
Material Safety Data Sheets (MSDS), 2
Melt Pressing, 192
Methyl Methacrylate,
 photopolymerization, 35
 suspension polymerization, 38
Microscope
 optical, 185, 186, 188
 polarizing, 186
Microscopy, optical, 185–197
 bright field, 188
 dark field, 189
 dynamics, 191
 embedding, 192
 interference, 189
 melt pressing, 192
 microtomy, 192
 morphology, 185
 phase contrast, 189
 refractive index, 190
 solvent casting, 192
 staining, 192
Microscopy, polarized light, 189
Microtomy, 192
Molecular Weight, 87, 88, 103, 133
 dynamic mechanical analysis, (DMA), 202
 gel permeation chromatography, (GPC), 88, 140–151
 end group analysis, 163, 166
 infrared spectroscopy, (IR), 103
 light scattering, 145, 152
 Mark–Houwink–Sakurada constants, 142
 Nuclear magnetic resonance, (NMR), 87–89
 number average, M_n, 88, 133, 134, 145, 148
 plasticizers, 202
 size average, M_z, 142
 size exclusion chromatography, (SEC), 140
 viscosity, dilute solution, 131
 weight average, M_w, 133, 134, 145, 146, 148
Molecular weight distribution, (MWD), 140, 145, 146

Morphology, 185
MSDS, see Material Safety Data Sheets
Mylar, 179

Nomex, 47
Notebook Records, 2
Nuclear Magnetic Resonance, (NMR), 83–97
  attached proton test, (APT), 89
  correlated spectroscopy, heteronuclear, (HETCOR), 89
  correlated spectroscopy, homonuclear, COSY, 89
  cross polarization magic angle spinning, CP/MAS, 90, 91
  distortionless enhancement of polarization transfer, (DEPT), 88, 89
  gyromagnetic ratio
  J-coupling, 85, 89
  kinetics, 92–94
  multiplicity, 86
  nuclear magnetic moment, μ, 83
  progressive saturation experiment, 90, 91
  solid state NMR, 89–91
  spectral editting, 88
  spin lattice relaxation, T_1, 84
  spin quantum number, I, 84
  spin spin relaxation, T_2, 84
  two-dimensional, 88
Nylon, 45f, 49
Nylon (6), 89, 165
Nylon (6,6), 165
Nylon (6,10) see Poly (hexamethylenesebacamide), 169
Nylon fibers
  thermogravimetric analysis, (TGA), 109

Operating Instructions, 3,4
Orientation, 103
  birefringence, 190
  degree of, 175, 176
  by infrared spectroscopy, 101
  molecular, 103
  uniaxial, 103
Osmometry, 163
Oxirane content, analysis for, 63
Oxycellulose, 165

Paraffin wax, 180
Particle Size Modification,
  in emulsion polymerization, 73

Parting agents, 29, 36
Percent solids, 31
Phenyl isocyanate, 165
Photopolymerization, 35
Poly(amide) 12, 45–51, 89, 188
  end group analysis, 164, 167, 168
  optically active, 47
  spherulite, 188
Poly(Bisphenol A-co-epichlorohydrin)
  glycidyl end capped, 92
Poly(butadiene), 86
Poly(1,4-butylene isophthalate),
  preparation, 59–60
Poly(caprolactam), 164
  mercapto groups, 165
Poly(carbonate), 86
Polyelectrolytes, 131, 133, 134
Poly(esters), 53–60, 86
  end group analysis, 164, 165
Poly(ethylene), 86, 87, 193
  branching, 87, 88
  dynamic mechanical analysis, (DMA), 204
  grafted, 165
  high density, 87
  irradiated, 165
  low density, 88
  microscopy, optical, 187
  solid state NMR, 90–91
  solution state NMR, 87–89
  X-ray diffraction, 179
Poly(ethylene glycol), grafted, 165
Poly(ethylene oxide), 144
Poly(ethylene terephthalate), PET, 131, 203
  dynamic mechanical analysis, (DMA), 204
  End group analysis, 165
  X-ray diffraction, 179
Poly(hexamethylenesebacamide),
  preparation of, 49
  end group analysis, 169
Polymers
  amorphous, 122, 185
  applications, 54
  branched, 134, 143
  chain branching, 140
  condensation, 163, 164
  crystalline, 122, 193, 202
  hydrodynamic volume, 141–143
  semicrystalline, 122, 173
  microscopy, polarized light, 189
Polymerization, 86
  addition, 86
  condensation, 86

Poly(methyl acrylate), 110, 111
   end group analysis, 165
Poly(methyl methacrylate), 82, 92–94, 144
   end group analysis, 165
   thermogravimetric analysis, (TGA), 110
Poly(methyl-1-pentene), 88
Poly(olefin), 86
Poly (oxymethylene), 193
Poly(phenylene oxide), 86, 202
Polysaccharide, 144
Poly(styrene), *see* Styrene
Poly(styrene), 86, 135, 140, 144, 146
Poly(styrene-co-maleic anhydride), SMA, 140
Poly(tetramethylene glycol), 167
Poly(urethene), 86
Poly(vinyl acetate) *see also* vinyl acetate
Poly(vinyl acetate)
   alcoholysis, 77–80
   crosslinking, evidence for, 77–78
   hydrolysis, 77
   preparation, 77–80
   as suspending agents, 31
Poly(vinyl butyral), 109
Poly(vinyl chloride), 86, 102
   dehydrogenated, 165
   dehydrohalogenation, 111
   differential scanning calorimetry, (DSC), 124
   dynamic mechanical analysis, (DMA), 204
   thermogravimetric analysis, (TGA), 112
Potassium bromate, 165
Potassium periodate, 165

Rayleigh factor, R_θ, 145
Rayleigh line broadening, 152
Redox Emulsion Polymerization, 41
Redox initiation systems, 32
Refractive Index, 123, 190
Repeat unit, 173
Rubber, 110
   thermogravimetric analysis, (TGA), 110

Safety Precautions, general, 2
Scale of Operations, 1
Schotten–Baumann reaction, 56
Seeded emulsion terpolymerization, 73–76
Sensitizers, Photo, 35
Shrinkage upon polymerization, 30

Size Exclusion Chromatography, (SEC), 140
Sodium fluoride, 178, 179, 181
   X-ray diffraction, 178, 179, 181
Sodium lauryl sulfate, 42
Sodium persulfate/sodium metabisulfite system, 42
Specific heat, c_p, 122, 124
Spherulites, 185, 186, 190
Stabilizers, 203
Stannic chloride, 24
Stereoregularity, 140
Styrene
   anionic polymerization, 17ff
   bulk polymerization, 10–12
   cationic polymerization, 22ff
   emulsion particle size, 14–15
   emulsion polymerization, 14
   free radical polymerization, 9ff
   heat of polymerization, 10
   polymerization, 7–25
   polymerization initiator, 9
   polymerization mechanism, 9–10, 18–19
Styrene-butadiene copolymers, 165
Surfactants, 31
Suspending agents, 31, 39–40
   poly(vinyl alcohol), 31
   sodium poly(methacrylate), 39
Suspension polymerization
   of acrylic esters, 27, 30
   of methyl methacrylate, 38

Tacticity, 86
   atactic, 86
   isotactic, 86
   stereoeqularity, 86
   syndiotactic, 86
Temperature, reaction, 28
   crystallization, T_c, 120, 123
   decomposition, 120
   degradation
   glass transition, T_g, 120, 123, 202
   melt, T_m, 120, 123, 124
Tensile strength, 202
Terpolymerization
   by emulsion process, 73
Tetramethylsilane, TMS, 84, 92
Thermal analysis, 90
Thermal degradation, 103
   infrared spectroscopy, (IR), 103
Thermal gravimetric analysis, (TGA), 108
Thermal initiators, 32

Thermogravimetric analysis, (TGA), 108–119
  additive, 108, 109
  adsorption, 108
  antioxidant, 108
  composition, 108, 109
  decomposition, 108, 109
  degradation, thermal, 108, 110
  depolymerization, 110, 111
  desorption, 108
  evaporation, 108
  flame retardancy, 108
  oxidation, 108
  qualitative analysis, 108, 109
  quantitative analysis, 108
  sublimation, 108
  zip length, 110
Transparency, 123
Tromsdorff effect, 29

Ubbelholde viscometer, 135, 136
UV radiation, 35, 36

Vinyl acetate *see also* poly(vinyl acetate)
  emulsion polymerization, 71, 73–76
  physical properties, 72
  polymerization of, 71–76
  suspension polymerization, 71
  in terpolymerization, 73–76
Vinyl neodecanoate
  in emulsion terpolymerization, 73–76
Vinyl polymers, 164
  end group analysis, 164
Viscosity, dilute solution, 131–139
  coefficient of viscosity, η, 131
  flexible chains, 133
  Huggins constant, k', 132
  Huggins equation, 132
  inherent viscosity, 131
  intrinsic viscosity, 131, 142, 146
  Kraemer constant, k'', 132
  Mark–Houwink equation, 133, 134
  molecular weight, 131, 133

Waste Disposal, 2

X-ray Analysis, 90
X-ray Diffraction, (XRD), 173–184
  amorphous polymers, 175, 178
  Braggs's Law, 175, 176, 181, 183
  crystalline polymers, 176, 178
  crystallinity, 173
  Ewald sphere, 174, 175
  fiber, 176, 177
  helices, 173
  Hull–Davey chart, 182
  lamellae, 173
  Miller Indices, 181, 182
  orientation, 175
  periodicities, 173
  powder diffraction, 176
  repeat unit, 173
  small angle scattering, (SAXS), 173, 179
  soft radiation, 178
  stereoregularity, 175
  unit cell, 173
  wide angle scattering, (WAXS) 173, 179

Zip length *see* thermogravimetric analysis, (TGA), 110